Human Impact on the Environment:
Ancient Roots, Current Challenges

EDITED BY

Judith E. Jacobsen
and
John Firor

Westview Press
BOULDER • SAN FRANCISCO • OXFORD

This Westview softcover edition is printed on acid-free paper and bound in library-quality, coated covers that carry the highest rating of the National Association of State Textbook Administrators, in consultation with the Association of American Publishers and the Book Manufacturers' Institute.

All rights reserved. No part of this publication may be reproduced or transmitted in any form or by any means, electronic or mechanical, including photocopy, recording, or any information storage and retrieval system, without permission in writing from the publisher.

Copyright © 1992 by Westview Press, Inc.

Published in 1992 in the United States of America by Westview Press, Inc., 5500 Central Avenue, Boulder, Colorado 80301-2877, and in the United Kingdom by Westview Press, 36 Lonsdale Road, Summertown, Oxford OX2 7EW

A CIP catalog record for this book is available from the Library of Congress.
ISBN 0-8133-8550-4

Printed and bound in the United States of America

 The paper used in this publication meets the requirements of the American National Standard for Permanence of Paper for Printed Library Materials Z39.48-1984.

10 9 8 7 6 5 4 3 2 1

304.28 SAC

Human Impact on the Environment:
Ancient Roots, Current Challenges

Published in Cooperation with Rice University and
Scientia, an Interdisciplinary Group at Rice University, Houston, Texas

Contents

List of Figures vii
Acknowledgments ix

PART ONE
Introduction 1

PART TWO
From Small Groups to Large:
The Impact of Hunting, Farming, and Cities 11

1 The Impact of Early People on the Environment:
   The Case of Large Mammal Extinctions, *Richard G. Klein* 13

2 The Impact of Food Production: Short-Term Strategies
   and Long-Term Consequences, *Charles L. Redman* 35

3 The Epidemiology of Civilization, *Mark N. Cohen* 51

PART THREE
The Industrial Era:
New Societies, New Technologies, New Problems 71

4 The Revolution in the Family and the World We
   Have Made, *David Levine* 73

5 Pollution and the Emergence of Industrial America,
   Martin V. Melosi 91

6   Exhaustibility of British Coal in Long-Run Perspective,
    G. N. von Tunzelmann   115

PART FOUR
The Environment Goes Global:
Issues of the Late Twentieth Century   141

7   Global Climate Change, *John Firor*   143

8   Global Water Resources: The Coming Crises,
    Peter H. Gleick   161

9   Tropical Forests and Human Society, *James D. Nations* 171

PART FIVE
Designing the Future:
Coping with the Crises   181

10 Creating an International Process to Address
    Greenhouse Warming, *William A. Nitze*   183

11 Human Impacts, *C. M. Hudspeth*   197

12 African Search for Solutions, *Thomas R. Odhiambo*   201

13 Transitions to a Sustainable Society, *James Gustave Speth* 207

About the Contributors   215
Index   217

Figures

1.1	Percentage of totally extinct larger mammal genera in successive time intervals in four regions	20
6.1	Quantity of coal mined annually in relation to cost per ton, 1981-1982	120
6.2	Quantity of coal mined annually relative to unit cost in 1918	121
6.3	Price per ton of coal sold in the North East coal field, 1837	122
6.4	Relative cost of mining coal 1868-1989, deflated by output price index	123
6.5	Relative supply price of coal at the pithead, 1700-1989, deflated by output prices	124
6.6	Relative supply price of coal at the pithead, 1700-1989, deflated by average wage rates	125
6.7	Relative demand price for coal at point of consumption, 1700-1989, deflated by retail prices	126
6.8	Patents issued in each five-year interval for access to coal	130
6.9	Patents issued in each five-year interval related to using coal	131
6.10	Patents issued in each five-year interval related to getting coal	132

6.11 Growth of coal output, 1700-1989   133

7.1 Concentrations of two infrared-trapping gases in the atmosphere during the past 10,000 years as deduced from measurements made on air trapped in polar ice and modern air samples   149

7.2 Surface temperatures during the past 1,000 years   154

Acknowledgments

The original impetus for this volume came from a symposium held at Rice University in April 1991 on the subject of Human Impact on the Environment. The idea of a conference on this subject arose within Scientia, an interdisciplinary group of members of the Rice faculty in specialties ranging from astronomy and anthropology to engineering, English, and history. This group was impressed with the fact that the assumption, common today, that all of the current modifications of the environment are "unnatural" implies that humans in earlier times did not damage the environment while improving their lot. Yet specific examples are known in which environmental damage occurred in very early civilizations. By organizing a conference that brought together specialists covering everything from relevant experiences revealed in the historic and pre-historic record to individuals engaged in understanding and resolving today's problems, Scientia hoped to foster new understanding of the relationship of people with their environment.

The conference was organized by Professor Don H. Johnson and was convened by Professor C. Robert O'Dell, both of Rice University. A number of other Rice faculty also participated in the planning, in the meeting itself, or both. These included Professors Patricia M. Bass, James B. Blackburn, Arthur Few, Tamara Ledley, George Marcus, Susan Keech McIntosh, E. S. A. Odhiambo, Ronald L. Sass, Albert Van Helden, and Martin Wiener; Deans Michael Carroll and Allen Matusow; Provost Emeritus William Gordon; and President George Rupp.

Except in two cases, the authors of the chapters of this volume are among those who gave presentations at the meeting. Experts in addition to these authors and the Rice faculty listed above also participated in the conference, including Kenneth L. Brown, University of Houston; Ralph Cicerone, University of California, Irvine; Roberto Pereira da Cunha, National Space Research Institute, São José dos Campos, Brazil; and John H. Gibbons, Office of Technology Assessment, U.S. Congress. Scientia had the invaluable support of Medaris and C. M. Hudspeth in transforming the idea of the conference into reality. All of these persons contributed in a variety of ways to this book, which grew out of their combined efforts and was made possible by them.

In the preparation of this volume, the editors had the very helpful advice and assistance of Ellen McCarthy and Kellie Masterson of Westview Press. Justin Kitsutaka and his staff at the National Center for Atmospheric Research prepared the figures, and Sheri Harms prepared the manuscript for reproduction.

Judith E. Jacobsen
John Firor

PART ONE

Introduction

This volume brings together a remarkably diverse group of scholars and writers. The work of paleontologists, anthropologists, historians, economists, hydrologists, biologists, physicists, and lawyers is found within its covers. Why? Each of them is addressing what many consider to be the most pressing issue of our time: the impact that human beings have on their environment, the physical and ecological world.

The works in this book consider that relationship in the longest possible sweep of history. The three chapters in Part Two, all written by anthropologists, chart the earliest impacts that humans had on the environment and on themselves--some the result of consciously-undertaken efforts to improve human life in the short term. The two historians and the economist writing in Part Three turn to particular implications of the industrial revolution: the impact on the family, the rise of efforts to deal publicly with the effluents of industrial activity, and the interaction of economics and politics in consideration of the exhaustibility of a mineral resource. In Part Four, a physicist, a hydrologist, and a biologist turn their attention to three dimensions of environmental problems that face us in the late twentieth century. They are the global greenhouse effect, issues surrounding global water resources, and tropical deforestation and the species loss that accompanies it. The last part considers strategies for coping with the crises faced today. Appreciating ancient, long-used indigenous methods of meeting human needs and the insight of a feminist perspective are discussed, as are the importance of curbing population growth, promoting energy and materials efficiency, and creating institutional arrangements to facilitate cooperation in international environmental problems. The precise dimensions of an international agreement to curb the emission of greenhouse gases are also suggested. Each of the chapters is discussed in greater detail below.

In Chapter 1, Richard G. Klein gives us the earliest possible example of human impact on the environment by analyzing the evidence of the effect that early humans had on mammal extinctions. He concludes that the evidence does not allow the conclusion that the earliest humans--premodern, or not fully modern humans--had a significant hand in important mammal extinctions. But the evidence is consistent with the conclusion that fully modern humans indeed contributed to mammal extinctions beginning approximately 100,000 to 200,000 years ago. The paucity and ambiguity of the fossil record from hundreds of thousands of years ago of course allows no absolutely confident conclusion. And usually animal extinctions occurred during the times that climate and environment were undergoing extreme changes, such as during the last ice age. But many of the species that went extinct during the ice age, after fully modern humans emerged, had survived earlier ice ages. Thus Klein concludes cautiously that humans beings had an early hand in contributing to large mammal extinctions. The human actions involved--hunting and competing for food--all occurred on the local scale; coordinated action on a larger scale is unlikely. But the impact, in the end, was global, as species of animals disappeared from the earth forever.

It is not possible to reconstruct what went on in the minds of the hunting and foraging bands living through the last ice age, or even to deduce general strategies that the hunters and foragers followed. But once humans began to plant crops and live in towns, it is possible to infer more about their lives. Charles L. Redman draws a number of important inferences in Chapter 2, inferences that are remarkably apt today. Analyzing three early urban civilizations, in Greece, what is modern Iraq, and the American Southwest, he is able to show how short-term strategies for maximizing food production had consequences over the longer term that were capable of destroying the civilization that the short-term strategies made possible.

Mark N. Cohen, in Chapter 3, presents some surprising facts about one long-term consequence of the short-term strategy of creating complex civilizations concentrated in cities: disease. He shows that contrary to widely-held assumptions, the health of human beings did not improve as we moved from hunting and foraging in small bands to living sophisticated lives in densely-populated cities, at least before the development of the germ theory of disease and antibiotics. In fact, urban life is a perfect setting for contagious infections. Furthermore, Cohen shows that the move from hunting and foraging to farming does not necessarily improve nutrition. And the less adequate food supply adds to the susceptibility to infection. As with the bargain to intensify agriculture with irrigation, which has Faustian elements, living in cities has carried with it the burden of disease. The plus side of the bargain was that the collective gained, the empire grew, the civilization flourished--even as the individual suffered.

Cohen concludes with a speculation that some of the ability of crowded civilizations to conquer less advanced peoples arose from this burden of disease. The less advanced--and more scattered--groups did not acquire the disease immunities carried by the citizens of cities, so they suffered crippling epidemics when they came in contact with "civilized" armies.

Redman's and Cohen's work has implications for our lives today. Neither recommends that we abandon civilization, or that we even regret that it evolved. Rather, it is well worth the effort to take a close, analytical look at things offered as "improvements" that promise short-term gains. Close examination may reveal long-term disadvantages and the tradeoffs required. Of particular importance may be the different scales at which the advantages and the disadvantages accrue. While governments and large businesses may benefit from technologies requiring large-scale manufacturing, the use of toxic or radioactive substances, the mining of minerals, or the drastic change in land-use patterns, individuals and communities may reap the disadvantages, in damaged health, fear of harm, or restricted opportunities.

Many works catalogue the nature of the Industrial Revolution--the rise in energy use, materials produced, and waste generated, for example, and the dramatic urbanization and population growth involved. Here, three pieces of that larger picture are dealt with in detail.

In Chapter 4, David Levine draws on a large and diverse literature to make several important points about changes in the family during the Industrial Revolution. Some of those changes are relatively well known: fertility fell dramatically and families moved from the countryside to the city, from a farming economy to a wage-earning one. Other changes discussed by Levine are less well known: that the behavior of individual family members became increasingly the subject of public and even state involvement; that the wife and mother in the family was assigned the task of enforcing order and cleanliness in the household; and that schools trained the good citizens of the future, inculcating the values of order, cleanliness, and low fertility. Perhaps the most important implication of Levine's study for the environmental problems of today is this: a close analysis of history shows that events do not "just happen" to people. Instead, the world is partly what we make it. How a particular set of circumstances or crises in history gives way to the next is the result of the interaction of large-scale phenomena, such as economics and politics, with individuals. It is useful to remember this when one begins to feel despair about solving environmental problems--individuals are an important piece of the picture.

In his history of public efforts to curb the sorts of industrial pollution produced by the energy sector, Martin V. Melosi in Chapter 5 reveals the roots of several biases in current efforts to protect the environment. He shows, for example, that pollution that harms human health is more likely to be cleaned up or prevented than pollution that is merely unsightly or damages only non-human parts of the environment--or cannot be shown to harm human health in the short term. He also shows that efforts to curb pollution have usually been based on the notion that pollution is the result of wastefulness in industrial activity, not that it is an intrinsic feature of industrial activity.

Chapter 6 traces the history of British coal over three centuries--costs of production, innovation in the industry, and fears of resource exhaustion. But perhaps the most important idea in G. N. von Tunzelmann's chapter is about the relationship between economics and politics. He shows that while experts have for a century analyzed whether British coal shows any economic signs of running out, in fact the question is intertwined with issues of the treatment of labor in the industry. Thus what seems to be a physical or economic question--is British coal running out?--is really a political one--how do we wish to treat coal miners? Von Tunzelmann's analysis has implications for environmental issues of today. We may ask, for example, whether it is "economically feasible" to improve energy efficiency in order to reduce the environmental impact of fuel use. But we find that this question is impossible to answer unless one also asks what kind of world we want to see in the future. This question--whether a consensus vision of a desirable future exists or is even possible--also surrounds decisions that today are being driven by new human impacts on the earth, those with a truly global dimension.

Part Three presents discussions of three important environmental problems we face today. John Firor explains the scientific basis for concerns that the climate may heat and otherwise rapidly change in Chapter 7. He discusses the "natural greenhouse effect," created by a number of gases in the atmosphere, among them carbon dioxide, methane, and water vapor. He then shows that those gases released by industrial civilizations are on the rise and reaching concentrations beyond those that occur naturally. Firor then describes the use of mathematical models of the climate system to project the warming of the atmosphere likely to occur if the concentrations of greenhouse gases continue to rise. While different models yield results different in some ways, it now appears probable that if societies continue to emit greenhouse gases as they have for the past several decades, the climate will change in many ways. Some of these changes are today unpredictable, and may remain so, but the global change--characterized by the estimated warming of 0.3 degrees Celsius per decade in the global average surface temperature--seems likely. While this warming sounds small when compared with the temperature extremes that each person has experienced, it is large for a change in the global average temperature, and perhaps more important, it is five times faster than any sustained change experienced by human civilizations since their beginning. More serious is the realization that this projected climate change is not an isolated phenomenon, to be solved in isolation, but part of the total complex of human impacts described in this volume and capable of solution only as part of a much more comprehensive drive to reduce all human impacts on the physical and biological world.

In his chapter on global water resources, Peter H. Gleick argues that water supplies and quality, so critical to so many human activities, are likely to become more and more constrained. As populations continue to grow they will require more food, and more food means more irrigated agriculture. As climates warm and evaporation increases, streamflows can diminish and water supplies become less certain. At the same time, many human activities tend directly to reduce water quality. Thus too many forces are converging on this most essential natural resource for us to be sanguine about its future.

In Chapter 9, James D. Nations tells the compelling story of tropical deforestation and the loss of species that it always entails. Focusing on Central America, he describes the economic forces that produce deforestation and the pattern that deforestation usually follows, from road-building to colonization to inappropriate agricultural practices. He offers a remedy for this global problem, a remedy that goes beyond merely forbidding forest clearing, if that were possible. He builds on the experience of indigenous peoples and migrants to forests who, instead of cutting them down for some other use, harvest the fruits of the forest year after year, earning income from trees indefinitely rather than destroying the capital that trees represent. He cites the palm cutters of Guatemala and the tagua harvesters of Ecuador. He also cites the history of forest products that have contributed enormously to human life on earth, from the major food crops to luxuries like coffee, bananas, and chocolate. His vision for saving the tropical forests involves, significantly, cultivating an economic return from the forests and drawing on the knowledge that indigenous peoples have of forest products and their uses.

The first chapter in the last section of the book, on strategies for dealing with the problems outlined in the third section, also argues for listening to the wisdom of indigenous peoples. Thomas R. Odhiambo cites examples from Africa of economic activities carried on for centuries without harm to the land--original examples of the "sustainable development" so prized by environmentalists and development experts today. In particular, the Wakara of Uganda and the Wachagga of Tanzania are given as examples of indigenous sustainable economies. Both groups have developed mixed, intensive systems that support their populations but do not do so at the expense of the environment, even over long periods of time.

C. M. Hudspeth, in Chapter 11, makes two points about our search for solutions to the environmental dilemmas of our times. First, he argues that rapid growth of the human population underlies all environmental problems and that any vision of a future in harmony with the environment must involve stabilizing world population growth. Second, he makes a point not often made in mainstream publications on the environment: that is that the culture that has brought the world to the constellation of environmental crises so familiar to us today is one that values at its most fundamental level the male competition, exclusion, and dominance of nature and of others. The first cultures were not this way; evidence suggests that they were matriarchal, inclusive, and cooperative with nature and with others. Again, any vision of a future in harmony with nature, the implication is, might turn a receptive eye to feminist ways of thought and action.

In Chapter 12, William A. Nitze argues for an international process that would encourage the countries of the world to work together to reduce the risks of a severe climate change. He recommends a combination of "top-down" and "bottom-up" negotiations and a procedure that would allow the evolution of agreements as our knowledge of climate science, climate change impacts, and policy responses grows. Such an enterprise, which Nitze sees as nothing short of forging a new global economic development strategy, would be aided by significant changes in technology, economics, and politics, some of which are already underway. Those changes include a shift from resource- and energy-intensive to knowledge-intensive production; improvements in manufacturing processes; biotechnological progress; new methods of measuring income, from the individual to the national level, that reflect environmental and social factors; and new political processes that involve local participation, institutional reform, and international cooperation. The final chapter in this volume, written by James Gustave Speth, provides a framework for the kinds of changes required to create a world in which people live in long-term harmony with the environment. After describing briefly the set of problems identified as environmental, and pointing out that the decade of the 1990s is the crucial one for turning these problems around, he suggests that six transitions are in order. They are demographic, technological, economic, social, institutional, and intellectual. It is necessary, he argues, to halt population growth before the world's population doubles again, to exceed 10 billion people. Technologies must become energy and materials efficient and environmentally benign. The world's economies must move from depletion of the earth's capital--its resources--to living on income. Socially, the world's environmental and economic benefits must be shared more equitably by the world's peoples and nations. Governments must be arranged to make better and quicker decisions about the environment and to be receptive to the wisdom of local people and others affected by government decisions. And finally, knowledge and understanding of the environment and of the impact of human beings on it must become more widespread.

This volume is one small step toward that last goal.

PART TWO

From Small Groups to Large: The Impact of Hunting, Farming, and Cities

1

The Impact of Early People on the Environment: The Case of Large Mammal Extinctions

Richard G. Klein

Introduction

As the twentieth century draws to a close, no issue is arguably more obvious or more pressing than the capacity of the human species to alter the face of the globe. It is equally obvious that by chance or design, this capacity has grown; it was less potent in past decades or centuries. By extension, it is widely assumed that no such capacity existed prehistorically and that it was certainly absent in the period before the end of the last ice age about 10,000 years ago, when people everywhere made their living as stone age hunter-foragers. My purpose here is to summarize the scanty evidence for human impact on the environment in this earlier period and to ask specifically whether this impact was anything more than incidental. This goal is constrained by the nature of the evidence, which is limited by preservational circumstances almost entirely to animal bones. But those bones signal one of natural history's most dramatic events: the extinction of many large mammals.

The principal question is whether people contributed to these extinctions. This is not a new question (Grayson 1984), and even a brief examination of the voluminous literature, including two dedicated volumes (Martin and Wright 1967; Martin and Klein 1984), will show that it is far from resolved. To some specialists, above all Martin (1967, 1984, 1990), the evidence points to a nearly exclusive human role; to others, for example Graham (1990), Graham and Lundelius (1984), Guthrie (1984, 1990), and Lundelius (1988, 1989), it suggests that climatic and environmental change was probably far more important to the extinctions. My own position is intermediate.

It is my view that people cannot be clearly implicated in any of the extinctions that occurred prior to the evolution of fully modern humans roughly 50,000 years ago, primarily because these extinctions are neither well dated nor clearly correlated with major human evolutionary events. Also, premodern human populations were arguably too small and their hunting and foraging technology too limited to have a significant impact on other animals. The problem of dating extinctions does not disappear with the advent of fully modern humans, but it is less severe. The apparent correlation of many subsequent extinctions with major archeological events circumstantially supports a human role. Major environmental, particularly climatic, changes often coincide with these later extinctions. But on present evidence, the changes did not differ significantly from much earlier changes that occurred without extinctions. What was new was the presence of fully modern humans, who were both more numerous and more efficacious hunter-foragers than any of their non-modern predecessors.

Beginning with an overview of the major events in human evolution, the sections that follow will document these conclusions. Particular reliance will be placed on African evidence. It is important to stress in advance that the argument made here could be supported or contradicted by fresh data, particularly by new extinction dates or by new evidence on the evolution of human ability to hunt and forage. At the same time, prior experience suggests that significant new data will probably appear very slowly, and the issue may never be conclusively resolved. In a very real sense, the issue of the human role in animal extinction resembles a legal case in which jurors faced with incomplete, ambiguous, or even contradictory evidence must still make a judgment. Like a trial, this issue is very different from a well controlled scientific experiment that can produce nearly complete closure on a question.

Human Biological and Cultural Evolution

Like the extinctions issue it partly encompasses, the evolutionary history of our species is controversial. The reason is the meagerness of the data: sparse, mainly fragmentary fossils and enigmatic artifacts from sites whose age and function are often not well established. To some extent, the continuing controversy also reflects diverse explanatory perspectives, but a firmer, more complete database would almost certainly reduce this diversity. The outline I offer here is based on a much longer summary (Klein 1989a). Citations are mainly to sources that appeared after that summary went to press. It represents a reasonable, if admittedly debatable, interpretation of the available evidence, stressing those aspects of human evolution that are most likely to bear on the human ability to extinguish other species.

The fossil record and the distribution of our closest living relatives, the chimpanzee and the gorilla, fix the birthplace of humanity in tropical Africa. The fossil record of east Africa places the birth date before 4 million years ago (Johanson 1989; White 1984), while the degree of molecular divergence between people and the African apes brackets the actual time between 10 and 5 million years ago (Hasegawa et al. 1989; Holmes et al. 1989). During this time the east African climate was becoming generally drier; grasslands and savannas were thus expanding at the expense of forests and woodlands. Arguably, the emergence of humans was an adaptive response to the selection pressures imposed by this vegetational change. The most obvious aspect of this response was a restructuring of the leg and foot to allow habitual, terrestrial bipedal locomotion. Indeed, it is primarily in this characteristic that the earliest people are recognizably human. Their skulls remained remarkably apelike, particularly in size.

The sedimentary context of early human fossils and associated objects, especially animal bones, demonstrates that the earliest-known people lived primarily, if not exclusively, in grassy or lightly wooded habitats. How they lived remains obscure, largely because the known sites are places where bones accumulated naturally, mainly in or on the margins of ancient lakes or streams. Contemporaneous occupation or living sites may never be found, if, like chimpanzees and gorillas, very early people were peripatetic and did not accumulate much refuse in any one place. Moreover, their refuse would not likely survive to become archeological evidence, if they either did not use tools or if the ones they used were like those of chimpanzees, highly informal and generally made of perishable materials (Boesch and Boesch 1990).

In the absence of an archeological record, the behavior and ecology of the earliest people must be inferred by assuming that they were a kind of ground-living, bipedal ape sharing significant behavioral traits with chimpanzees. On this basis, it seems likely that they subsisted mainly on vegetal foods, for which they foraged each day. Aside from insects, and occasional small vertebrates which they opportunistically encountered, they probably obtained little animal protein. With limited facility to communicate, little or no technology, and a social organization that likely involved relatively little cooperation among individuals, they were doubtless highly vulnerable to predation. This may explain why their arms and hands retained apelike features that would have facilitated tree climbing for escape (Aiello and Dean 1990; Susman, et al. 1984). Overall, whether measured by their numbers or by their interactions with other species, they were probably an inconspicuous element of the contemporaneous large mammal fauna.

These beings continued to evolve, however, and between 3 million and 2.5 million years ago walking on two legs on the ground became successful enough to permit the development of two distinct subadaptations, represented by two sympatric human lineages. Judging by facial and dental morphology, one lineage was arguably specialized for masticating hard or grit-encrusted plant tissues removed from the ground (Grine 1988). It thus remained largely vegetarian. In contrast, the other, in the direct line of ancestry to ourselves, became omnivorous. It incorporated a larger amount of animal protein in its diet, probably with the help of stone tools, which have been recovered at several African sites dated between 2.5 and 1.8 million years ago (Harris 1983; Harris and Semaw 1989; Harris, et al. 1990). The appearance of stone tools is broadly correlated with an increase in brain size, which may have been both cause and effect of stone tool manufacture. Some sites where tools are especially abundant may have been base camps that signal a shift towards characteristically human social organization, and most have provided animal bones that probably represent food debris. Specialists are still debating whether the bones were acquired by hunting or some form of scavenging (Bunn and Kroll 1986; Potts 1988), and the issue may never be settled with the evidence on hand. However, some of the bones are damaged by stone tools, and they commonly come from creatures far larger than those eaten by chimpanzees. The implication is for meat eating that is much more typically human than apelike.

By roughly 1 million years ago, armed with yet larger brains and more sophisticated stone tools, the lineage ancestral to ourselves had spread throughout Africa, excepting only the extreme deserts and tropical rainforest. About the same time, the other, more strictly vegetarian lineage became extinct. The reason is not clear, but an inability to compete successfully with its evolving cousin is a possibility discussed below. Then, at a precise time that remains debatable, evolving humans colonized Eurasia. Presumably they moved first from northeast Africa to the Near East (southwestern Asia) and then northwestward to temperate

Europe and eastward to Java and China. The expansion into north temperate latitudes may have been facilitated by advances in the ability to obtain meat. In addition, a developing control over fire would appear essential, but concrete evidence is lacking. Vestiges of fire that could reflect either human or natural causes have been found at African sites antedating one million years ago (Brain and Sillen 1988; Clark and Harris 1985; James 1989), but the oldest traces commonly credited to people date from half a million years ago, at the famous Peking (Beijing) Man site in China.

The period between roughly 1 million years ago and about 130,000 years ago is especially hard to summarize. Most of the few relevant sites are poorly dated with respect to each other. However, the admittedly sparse fossil record can be read to suggest that there were three distinct human evolutionary trajectories, based in Europe, east Asia, and Africa. The European trajectory led eventually to the Neanderthals, whose fossil remains and archeological debris are well-known from sites dated between 130,000 and 40,000 years ago. The trajectory in east Asia produced people who were as archaic as the Neanderthals, but distinct from them in significant morphological details. In strong contrast, both the fossil record and the genetics of modern human populations suggest that the trajectory in Africa led to the development of anatomically modern people, probably sometime between 200,000 and 100,000 years ago (Stringer 1989, 1990).

After 130,000 years ago, people in all three areas appear to have controlled fire. In general they made more sophisticated stone artifacts than their predecessors. Probably they were also more effective hunter-foragers. The archeological record does not suggest that major behavioral differences among continents coincided with the apparent physical ones, however. Thus, the earliest moderns in Africa made essentially the same kinds of stone artifacts as the Neanderthals, and neither group seems to have recognized bone, shell, and related substances as materials that could be carved, polished, or otherwise shaped into tools. Perhaps partly because of this, neither group has left us any indisputable evidence for art, with its obvious implications for fully human cognitive and communicative facilities. A behavioral difference, prominently involving the development of art as well as important advances in more strictly utilitarian artifact manufacture, developed only after about 60,000 years ago. The archeological record remains mute on where this occurred, but Africa or its Near Eastern periphery seems the likely place. If this is so, the competitive advantage it conferred would explain how modern people were then able to spread rapidly at the expense of biologically and behaviorally more primitive humans. Beginning roughly 50,000 years ago, moderns not only quickly replaced the Neanderthals and similarly archaic people throughout the occupied world, but they also became the first humans to occupy northernmost Eurasia and to colonize the New Worlds of Australia and the Americas.

The evolution of modern human behavior, or perhaps more precisely, of the modern ability to use culture as an adaptive mechanism, radically altered the previous pattern of human evolution whereby physical and biological changes occurred more or less hand-in-hand at a relatively slow pace. In the 50,000 years or so years following the modern human diaspora, fundamental change in the human form essentially ceased, but behavioral or cultural change accelerated dramatically. These changes led eventually to the development of food-producing economies about 10,000 years ago (Sheratt 1980; Blumler and Byrne 1991) and from there to cities and other elements of "civilization" only 5,000 later (Oates 1980). It is the uniquely modern ability to employ culture that makes the human species a threat to so many others today. As suggested below, the destructive potential of this ability may be traced back to nearly the time of its emergence.

The Extinctions Record

A comprehensive understanding of extinctions must cover all creatures, but considerations of a human role traditionally focus on so-called large mammals, housecat-size or larger. This is partly because large mammals tend to dominate pertinent fossil assemblages. Also, they were probably the most important species in the lives of early people, whether the people hunted, scavenged, or simply sought to avoid predation. The traditional focus on large mammals is kept here.

An analysis of large mammal extinctions designed like a laboratory experiment would obviously begin with thorough sampling of the fossil record at close, well-dated intervals. Unfortunately, the vagaries of preservation and discovery preclude anything approximating complete sampling, and precise dating is often a problem, particularly at sites that do not provide suitable materials for radiometric methods. Unresolved or uncertain dates remain a major obstacle not only to the study of extinctions, but to all related geohistorical endeavors, including human paleontology. This is especially true in the interval between the lowest age detectable by conventional radiopotassium methods (between 1 million and 500,000 years ago) and the oldest specimens measurable by conventional radiocarbon methods (roughly 40,000 years ago). Reconstructing the extent and chronology of extinctions is complicated further by inevitable disagreements among specialists about the species of particular fossils and whether a species is truly extinct or simply ancestral to a later form.

Problems of sampling, dating, and species identification can never be fully overcome, but they can be minimized by concentrating on relatively broad time intervals in regions that have provided large, thoroughly described fossil samples. The problem of species confusion can also be partly circumvented by focusing on genera instead. Using these devices, a picture of large mammal extinctions and the possible role of humans on them emerges.

Figure 1.1 presents that picture in four different regions. This figure shows the proportion of now-extinct large mammal genera as this proportion is presently estimated for successive intervals beginning up to 5.5 million years ago and ending about 10,000 years ago. "Now-extinct" genera are ones that apparently terminated without issue at or before 10,000 years ago. The regions are southern Africa, the Near East, Europe, and North America. Other potentially suitable regions have been excluded not only because their data are less complete, but also because in some cases the data are less intelligible to specialists in other regions. The beginning date was chosen because it essentially encompasses the entire known record of human evolution. The 10,000 year cutoff is convenient not only because it represents the last time that people everywhere lived exclusively by hunting and foraging, but also because it marks the end of the last glaciation--the most recent of more than twenty glacial-to-interglacial transitions that have occurred since the initiation of such cycles roughly 2.5 million years ago (van Donk 1976; Prentice and Denton 1988; Williams, et al. 1988). It was thus a time of profound environmental change with potential implications for extinctions. It was notably also the only glacial-to-interglacial transition to postdate the evolution of fully modern people.

Not surprisingly, Figure 1.1 shows that within each region, the proportion of totally extinct genera declined through time. Another potentially more interesting pattern is that the greatest drop in the proportion of extinct genera varies in time from region to region. Most notably, it occurs much earlier in southern Africa than in North America. Since among the regions considered, southern Africa is certainly the one that people inhabited first and North America is the one that they colonized last, the broad pattern might suggest a link between apparent generic loss

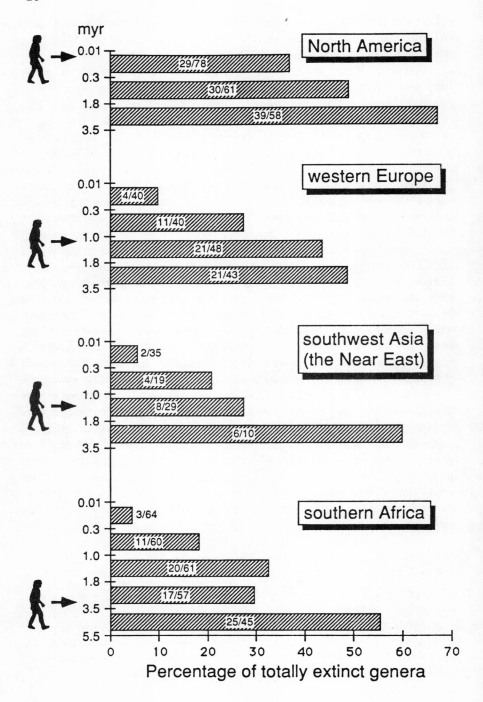

and human presence. This possibility is explored further in the next two sections, first for the extinctions that occurred before the evolution of fully modern people and then for those that occurred afterwards.

Extinctions and Premodern Humans

Martin (1967) was the first to marshal substantial evidence implicating premodern people in large mammal extinctions. His argument was based primarily on what he saw as a close temporal correspondence between extinctions and significant archeological events in different regions--events such as initial human habitation or colonization and major, subsequent artifactual changes or advances. He ignored the question of premodern hunting or predatory prowess, a very difficult subject to address.

Pertinent archeological sites are mainly places where artifacts and animal bones accumulated on the margins of ancient lakes or streams. In virtually all instances, animal deaths may have been largely natural, occurring at points where people and other creatures routinely came to drink or perhaps to shelter under shade trees. The duration of bone and artifact accumulation usually cannot be established. But in most instances, months, years, decades, or millennia are more probable than days or weeks. Thus bones and artifacts that now lie next to one another in the ground may actually have been deposited days, weeks, or months apart.

At some sites, bones damaged by stone tools indicate some human role. But often a carnivore role is at least equally implied by tooth marks or by fossilized carnivore feces (coprolites). Archeologists have not yet developed methods for isolating the nature and extent of human involvement. If it is fair to extrapolate backwards from the apparently limited hunting ability of the earliest modern people (Klein 1989b), it may be fair to say that all premoderns were too ineffectual to precipitate extinctions. But until direct evidence becomes available, the case must rest entirely on the kind of correlations that Martin proposed.

FIGURE 1.1 (left) Percentage of totally extinct larger mammal genera in successive time intervals in four regions. Numbers on each bar are the number of totally extinct genera divided by the total number of known genera. Time is presented in millions of years (myr) before present. Note that successive time intervals are of unequal length. Arrows indicate the approximate times when people first appeared in each region. *Source:* See References.

Figure 1.1 shows that in each of the four regions--southern Africa, the Near East, Europe, and North America--a broad correspondence emerges between the oldest traces of human occupation and a major decline in the proportion of now-extinct genera, that is, with a possible extinction event. But the actual situation is much more complex than the figure suggests. The problem is partly that in any given region people may actually have been present much earlier, especially in Europe, where some authorities continue to argue for a human presence at least 2 million years ago (Bonifay 1989; Bonifay, et al. 1989; Delson 1989), or a million years earlier than the date in Figure 1.1. Their case rests mainly on crudely flaked stones whose artifactual nature is--and may always be--debatable. If a date of 2 million years ago is ultimately confirmed, it would also imply equally early occupation of the Near East, which lies astride the most probable route to Europe.

Even if the timing of first human colonization or appearance could be firmly fixed, a bigger problem remains: to date the disappearance of each extinct genus, either absolutely or with respect to the disappearance of others. At the moment, no evidence suggests that the extinctions occurred in clusters or waves that could correspond to a major archeological or paleoanthropological event. Further, it is at least equally probable that extinctions occurred gradually over the long time intervals into which Figure 1.1 groups them. In this regard, it is relevant that significant large mammal extinctions occurred in each region long before people arrived. Arguably, the rate of extinction increased afterwards (Maglio 1975), but in each region the period after human arrival was also a time of unusually marked climatic and environmental change, during which glacial-to-interglacial fluctuations became generally more intense. Under the circumstances, climatic change and the accompanying reshuffling of biotic communities certainly constitute a plausible and adequate, if not conclusive, explanation for the known extinctions record.

This is not to say that premodern people did not sometimes precipitate extinctions, perhaps indirectly. And fresh data may yet demonstrate a potentially significant chronological coincidence between a major paleoanthropological event and one or more extinctions. This is perhaps particularly likely if the focus is limited to special aspects of the extinctions record, such as the tendency for carnivore extinctions to exceed those of herbivores in southern Africa between roughly 5.5 and 0.5 million years ago (Hendey 1974a; Klein 1984), or the termination of the second, more conservative (so-called "robust") human lineage that existed alongside our own until sometime between 1.2 and 0.7 million years ago (Klein 1988). We might ultimately find, for example, that important carnivore extinctions followed closely on a major human advance in the ability to obtain meat, or that the second hominid lineage disappeared at a time when its evolving sibling was rapidly encroaching on its resource base. In either case, it might be argued that premodern people precipitated extinctions indirectly, through niche absorption or competitive exclusion.

Extinctions and Fully Modern Humans

The possibility that early modern humans caused large mammal extinctions is far easier to investigate. The time interval involved--the last 200,000 or 100,000 years--is relatively recent, and the relevant fossil and archeological records are much more complete and better dated than the records for premodern people. The result is that both potentially relevant human behavior and the timing of extinctions are far better understood.

The key human behavioral observation concerns an apparent quantum advance that occurred sometime between 60,000 and 40,000 years ago (Mellars 1989). As mentioned before, prior to this time, all people, including the earliest anatomical moderns in Africa and their Neanderthal contemporaries in Europe, seemed not to have realized that bone, antler, ivory, and related substances could be carved, polished, or otherwise shaped into formal artifacts like those that archeologists call "points," "awls," "needles," and the like. Perhaps partly as a result, unlike their successors after 40,000 years ago, they left us no firm evidence for art, whether in the form of engravings, carvings, or paintings or in the form of beads, pendants, and other ornaments. Their stone tools were arguably more sophisticated than those made by their predecessors before 200,000 or 100,000 years ago, but overall, the tools comprised a smaller number of types than those made after 40,000 years ago, and the types themselves are more difficult to distinguish, suggesting they were produced from less precise or more vaguely defined mental templates. Perhaps most important, the artifact assemblages that antedate about 40,000 years ago commonly exhibit far less variability through time and space than later ones, whose remarkable spatial and temporal heterogeneity has allowed archeologists to recognize numerous distinctive artifact "cultures," even within relatively small regions or short time spans. In short, from the artifacts alone, we could conclude that human behavior changed qualitatively roughly 50,000 to 40,000 years ago and that it was only afterwards that people possessed the fully modern human ability to innovate in response to changing environmental or social conditions.

Other recoverable behavioral clues tell the same story. Thus, the earliest moderns and the Neanderthals buried their dead, but their known graves are all relatively simple; they were often just large enough to hold a body, and they never contain unequivocally special objects or "grave goods." More elaborate graves implying a burial ritual and associated religious beliefs appear only after about 50,000 years ago, in the same sites that have provided the earliest formal bone artifacts and art objects. Similarly, sites older than about 50,000 years sometimes contain types of stone that originated many kilometers away, but it is only in later sites that such "exotic" raw materials are truly abundant. The implication may be that the kinds of wide-ranging exchange systems or social networks that are so familiar to us ethnohistorically developed only after about 40,000 years ago.

In the same vein, people before this time knew the use of fire, but their hearths were far simpler than many that were made afterwards. This is particular true in Eurasia, where complex construction implies a detailed knowledge of how to maximize heat production. More effective fireplaces probably help explain why it was only after about 50,000 years ago that people first colonized the harshest, most continental parts of Eurasia in what is today European Russia and Siberia. Also important is that only sites from after that time have provided readily recognizable "ruins," demonstrating an ability to construct substantial, well-insulated dwellings.

In Europe, a relatively sharp increase in the number of sites per millennium suggests that people after 40,000 years ago significantly outnumbered their Neanderthal predecessors living under comparable climatic conditions. In southern Africa, a broadly contemporaneous population increase may be inferred from a significant decline in average tortoise and shellfish size in archeological sites. Tortoises and shellfish were undoubtedly collected rather than hunted, and their smaller size implies a substantial increase in collection pressure, most likely because the number of collectors had increased. Also in southern Africa, it was only after about 50,000 years ago that people apparently developed the ability to catch fish and flying birds routinely and to obtain buffalo, wild pigs, and other especially dangerous prey in rough proportion to their live abundance (Klein 1989b). Future research may show that similar advances in hunting and foraging competence also occurred elsewhere in Africa and in Eurasia at about the same time. It is surely no coincidence that only artifact assemblages postdating 50,000 years ago contain bone and stone artifacts that can be reasonably identified either as fishing or fowling gear or as parts of advanced hunting weapons. Perhaps the most significant innovation was the bow and arrow, whose presence in southern Africa at least 20,000 years ago is strongly suggested by stone and bone artifacts that are strikingly similar to parts of historically observed composite arrows.

Since anatomically modern humans apparently evolved in Africa much earlier than 50,000 years ago, it seems probable that the advance to fully modern behavior also occurred there, and the lag of behavior behind anatomy would explain why the earliest anatomically modern humans failed to supplant the Neanderthals and other archaic Eurasians early on. In both body and face, the Neanderthals may have been physiologically adapted to glacial cold in Europe (Trinkaus 1989). Without a behavioral advantage, early moderns may therefore have been unable to displace them. Unfortunately, the African archeological record provides no clues as to how and why fully modern behavior evolved when it did. Perhaps it was largely an adaptive response to the hyperarid conditions that gripped much of Africa in the middle of the Last Glaciation, around 50,000 years ago. If so, the same aridity has impeded archeological progress by reducing the archeological visibility of contemporaneous African populations. Sites that probably date between 60,000 and 40,000 years ago in Africa are almost unknown.

Regardless of exactly when, where, and how fully modern human behavior appeared, it does not seem to have precipitated any large mammal extinctions in either Africa or Eurasia. Excepting minor redistributions of species that are readily explained as responses to climatic fluctuations, neither continent experienced any dramatic change in large mammal communities until about 11,000 years ago. At that time, there were numerous local extinctions on both continents, including, for example, the disappearance of reindeer over much of Europe. Some permanent losses also occurred, including, in sub-Saharan Africa, two or three genera and three or four species in surviving genera (Klein 1984; Marean and Gifford-Gonzalez 1991), and in Eurasia, three genera (Kurtén 1968; Stuart 1982, 1983; Sutcliffe 1985). The African genera that disappeared were Pelorovis, a large buffalo, Megalotragus, a large relative of the living hartebeest, and less certainly because of dating problems, Metridiochoerus, a large relative of the living warthog. The European genera were Mammuthus, the mammoth, Coelodonta, the woolly rhinoceros, and Megaceros, the "giant" or Irish Elk. (The topmost bar for Europe in Figure 1.1 includes these three genera and a fourth, Dicerorhinus, a rhinoceros that apparently became extinct much earlier, perhaps at the beginning of the Last Glaciation, roughly 70,000 years ago.)

Because the period around 11,000 years ago was one of profound climatic change in the transition from the last glaciation to the present interglacial (or Holocene), it is tempting to explain the extinctions climatically. This is all the more so since the numbers and the ranges of the taxa that disappeared were almost certainly diminished by the kind of vegetational changes that took place. These changes include, for example, the replacement of steppe-tundra by forest over much of temperate Europe. However, the same taxa survived the comparable environmental change that occurred during previous glacial-to-interglacial transitions, including the one-before-last, roughly 128,000 years ago, when no large mammals appear to have become totally extinct in either Eurasia or Africa. The only way in which the last glacial-to-interglacial transition differed conspicuously from earlier ones was in the presence of much more sophisticated, fully modern hunter-foragers. It thus seems reasonable to suppose that people were involved, not necessarily by pursuing extinct species to the last individual, but simply by perturbing regional ecosystems, so that they emerged from flux somewhat differently than they would have otherwise.

The logic of this explanation, which combines environmental change and human actions, may be extended to the significantly more extensive extinctions that occurred in North America roughly 11,000 years ago. High-quality radiocarbon dates indicate that at least eight large mammal genera made their last appearance here at this time (Grayson 1988, 1989, 1991). These were primarily genera that appear to have been very widespread and common and whose demise is thus relatively easy to date. Future research may show than an additional twenty-seven genera, more rare and more geographically restricted, also disappeared in the same interval. The alternative is that they disappeared earlier; almost certainly none survived 10,000 years ago. Of the thirty-five genera that may have disappeared about 11,000 years ago, six survived on other continents, but 29 vanished completely and without replacement. The overall result was a dramatic impoverishment of the North American mammal fauna.

As in Africa and Eurasia, the period 12,000 to 10,000 years ago in North America was a time of sweeping climatic and environmental change. These changes probably reduced the numbers and ranges of many taxa, including many of the extinct ones. However, again as in Africa and Eurasia, the same taxa had survived earlier intervals of similar change, and the only conspicuously novel element was human. The people involved are commonly called Paleoindians. Controversy continues over how long they had been present (Grayson 1988; Lynch 1990; Morlan 1988). Most authorities do not place Paleoindians in North America before 14,000 years ago. But a date up to 40,000 years is conceivable, based on the earliest colonization of northeast Asia (Siberia), from which both genetic evidence and geographic logic suggests they must have come. Even if the earlier date is correct, the Paleoindians were still relative newcomers at the time of the extinctions, and their limited integration into North American ecosystems could explain why large mammal loss under radically changing climatic conditions was so much greater than in Eurasia or Africa.

The case for fully modern hunter-foragers as an ingredient in extinctions is clearly not exhaustive. It could be refuted by a variety of new evidence, indicating, for example, that the most recent glacial retreat was climatically unique or that many prominent North American genera became extinct before people arrived. Its overall plausibility could also be significantly affected by future research in Australia, where the archeological and fossil record suggests some striking parallels and differences with North America. The principal parallel is that a wave of extinctions, which reduced the number of Australian large mammal species by 40 to 70 percent (Murray 1984; Horton 1984), may have closely followed the first arrival of people. Like the Paleoindians, the first Australians appear to have been fully modern. The colonization of Australia probably required the fully modern ability to innovate--Australia could be reached only by crossing the large stretches of open sea that separated Australia from the east Asian mainland, even during glacial intervals when sea level was much lower.

The major difference to North America is that both the colonization of Australia and the extinctions that may have followed occurred much earlier, perhaps as early as 45,000 years ago (Jones 1989). This period is in the middle of the last glaciation, when climatic change was relatively limited, implying that relative to climate, early Australians may have played a much greater role in extinctions than their counterparts in North America, Eurasia, or Africa. Thus, Australia could provide the clearest example of major human impact on the environment prior to 10,000 years ago. For the moment, however, the case remains open, because the Australian extinctions are imperfectly dated and many may antedate human occupation. So far, only a handful of extinct forms have been found at sites that demonstrably postdate human arrival (Horton 1984; Jones 1989). In sum, Australia could prove more relevant to the extinctions debate than any other continent, but only if a denser and much more firmly dated fossil record can be obtained.

Summary and Conclusions

Did people before the end of the last ice age affect their environment? The answer is surely yes, just as it is certain that their environment affected them. Was their effect ever great enough to produce profound, even irreversible environmental changes? Here the answer is far less certain, mainly because the surviving evidence is so sparse and equivocal. Arguably, early humans were too rare and too limited technologically to have much impact, particularly before their initial colonization of Eurasia, 1 million years ago or shortly before. However, the animal remains that comprise the most abundant and conspicuous direct evidence have proven remarkably difficult to interpret. And they have not yet permitted a forthright assessment of hunting and foraging capabilities before the emergence of anatomically modern humans within the last 200,000 to 100,000 years. Significant human influence on the environment before 200,000 to 100,000 years ago might be inferred from chronological correspondences between major human evolutionary events and the extinction of other species, but the fossil record remains too scanty and too poorly dated to determine if such correspondences exist.

Potential or likely human impact on the environment after modern humans evolved is easier to assess, mainly because the archeological and fossil record is much more complete and more firmly dated. Particularly notable is a qualitative advance in human cultural capabilities that occurred between 60,000 and 40,000 years ago and that led to increases in human population densities, an expansion of the human ecumene, and breakthroughs in hunting-foraging proficiency. All human populations after about 40,000 years ago appear to have possessed the completely modern ability to employ culture as an adaptive mechanism, and in the succeeding interval, cultural evolution accelerated dramatically while fundamental human morphological change ceased. By 10,000 years ago, relatively sophisticated hunter-foragers were inhabiting virtually every corner of the globe, and it is readily conceivable that they were at least partly responsible for the many large mammal extinctions that occurred at this time.

The interval between 12,000 and 10,000 years ago was also a period of profound climatic and environmental change in the transition from the last glaciation to the present interglacial, and this almost certainly reduced both the ranges and numbers of many of the species that vanished. However, the same species had survived earlier glacial-to-interglacial transitions, and the only factor that clearly differentiated the last transition was the presence of more advanced hunter-foragers. The recency of the last glacial/interglacial transition means that new evidence on the accompanying extinctions event is relatively easy to obtain, and many researchers are seeking it out. As a result, the case for or against human involvement should be much stronger a few years hence.

Acknowledgments

I thank D.K. Grayson for criticisms of a preliminary draft and the National Science Foundation for supporting my own research on large mammal extinctions.

References

Data sources for Figure 1.1: For southern Africa, Brain (1981), Brain et al. (1988), Hendey (1974a, 1974b), and Klein (1984); for the Near East, Tchernov (1984, 1987, 1988); for Europe, Azzaroli et al. (1988), Gibbard et al. (1991), Kahlke (1975), Kurtén (1968), and Stuart (1982); for North America, Kurtén and Anderson (1980), Grayson (1989, 1991), and Martin (1984).

Aiello, L. C. and M. C. Dean. 1990. *An Introduction to Human Evolutionary Anatomy.* London: Academic Press.
Azzaroli, A., C. de Guili, G. Ficcarelli, and D. Torre. 1988. Late Pliocene to early mid-Pleistocene mammals in Eurasia: Faunal succession and dispersal events. *Paleogeography, Paleoclimatology, Paleoecology* 66, 77-100.
Blumler, M. A., and R. Byrne. 1991. The ecological genetics of domestication and the origins of agriculture. *Current Anthropology* 32, 23-54.
Boesch, C. and H. Boesch. 1990. Tool use and tool making in wild chimpanzees. *Folia Primatologica* 54, 86-99.
Bonifay, E. 1989. Un site du très ancien paléolithique de plus de 2 M. a. dans le Massif Central français: Saint-Eble-Le Coupet (Haute-Loire). *Comptes Rendus Hebdomamaires des Seances de l'Academie des Sciences,* Paris 308, Série II, 1567-1570.
Bonifay, E., A. Consigny, and R. Liabeuf. 1989. Contribution du Massif Central français à la connaissance des premiers peuplements préhistoriques de l'Europe. Comptes Rendus Hebdomadaires des Séances de l'Académie des Sciences, Paris, 308, Série II, 1491-1496.
Brain, C. K. 1981. *The Hunters or the Hunted? An Introduction to African Cave Taphonomy.* Chicago: University of Chicago Press.
Brain, C. K., C. S. Churcher, J. D. Clark, F. E. Grine, P. Shipman, R. L. Susman, A. Turner, and V. Watson. 1988. New evidence of early hominids, their culture and environment from the Swartkrans Cave, South Africa. *South African Journal of Science* 84, 828-835.

Brain, C. K., and A. Sillen. 1988. Evidence from Swartkrans Cave for the earliest use of fire. *Nature* 336, 464-466.

Bunn, H. T. and E. Kroll. 1986. Systematic butchery by Plio-Pleistocene hominids at Olduvai Gorge, Tanzania. *Current Anthropology* 27, 431-452.

Clark, J. D. and J. W. K. Harris. 1985. Fire and its roles in early hominid lifeways. *The African Archaeological Review* 3, 3-27.

Delson, E. 1989. Oldest Eurasian stone tools. *Nature* 340, 96.

Gibbard, P. L., R. G. West, W. H. Zagwijn, P. S. Balson, A. W. Burger, B. M. Funnell, D. H. Jeffery, J. de Jong, T. van Kolfschoten, A. M. Lister, T. Meijer, P. E. P. Norton, R. C. Preece, J. Rose, A. J. Stuart, C. A. Whiteman, and J. A. Zalaisiewicz. 1991. Early and early middle Pleistocene correlations in the southern North Sea Basin. *Quaternary Science Reviews* 10, 23-52.

Graham, R. W. 1990. Evolution of new ecosystems at the end of the Pleistocene. In *Megafauna and Man: Discovery of America's Heartland*, ed. L. D. Agenbroad, et al. 54-60. The Mammoth Site of Hot Springs Scientific Papers 1.

Graham, R. W., and E. L. Lundelius. 1984. Coevolutionary disequilibrium and Pleistocene extinctions. In *Quaternary Extinctions*, ed. P. S. Martin and R. G. Klein, 223-249. Tucson: University of Arizona Press.

Grayson, D. K. 1984. Nineteenth-Century explanations of Pleistocene extinctions: A review and analysis. In *Quaternary Extinctions*, ed, P. S. Martin and R. G. Klien, 5-39. Tucson: University of Arizona Press.

_____. 1988. Perspectives on the archaeology of the first Americans. In *Americans Before Columbus*, ed. R. C. Carlisle, 107-123. Ethnology Monographs 12, Department of Anthropology, University of Pittsburgh, Pittsburgh.

_____. 1989. The chronology of North American late Pleistocene extinctions. *Journal of Archaeological Science* 16, 153-165.

_____. 1991. Late Pleistocene mammalian extinctions in North America: Taxonomy, chronology, and explanations. *Journal of World Prehistory* 5, in press.

Grine, F. E. 1988. Evolutionary history of the 'robust' australopithecines: A summary and historical perspective. In *Evolutionary History of the Robust Australopithecines*, ed. F.E. Grine, 509-520. New York: Aldine de Gruyter.

Guthrie, R. D. 1984. Mosaics, allelochemics, and nutrients: An ecological theory of late Pleistocene megafaunal extinctions. In *Quaternary Extinctions*, ed. P. S. Martin and R. G. Klein, 259-298. Tucson: University of Arizona Press.

_____. 1990. Late Pleistocene faunal revolution: A new perspective on the extinction debate. In *Megafauna and Man: Discovery of America's Heartland*, ed. L.D. Agenbroad, J. I. Mead, and L. W. Nelson. 42-53. The Mammoth Site of Hot Springs Scientific Papers No. 1.

Harris, J. W. K. 1983. Cultural beginnings: Plio-Pleistocene archaeological occurrences from the Afar, Ethiopia. *The African Archaeological Review* 1, 3-31.

Harris, J. W. K., and Semaw, S. 1989. Further archaeological studies at the Gona River, Hadar, Ethiopia. *Nyame Akuma* 31, 19-21.

Harris, J. W. K., P. J. Williamson, P. J. Morris, J. de Heinzelin, J. Verniers, D. Helgren, R. V. Bellomo, G. Laden, T. W. Spang, K. Stewart, and M. J. Tappen. 1990. Archaeology of the Lusso Beds. In *Evolution of Environments and Hominidae in the African Western Rift Valley*, ed. N.T. Boaz. 237-272. Martinsville, Virginia: Virginia Museum of Natural History.

Hasegawa, M., H. Kishino, and T. Yano. 1989. Estimation of branching dates among primates by molecular clocks of nuclear DNA which slowed down in Hominoidea. *Journal of Human Evolution* 318, 461-476.

Hendey, Q. B. 1974a. The late Cenozoic Carnivora of the South-Western Cape Province. *Annals of the South African Museum* 63, 1-369.

———. 1974b. Faunal dating of the Late Cenozoic of southern Africa, with special reference to the Carnivora. *Quaternary Research* 4, 149-161.

Holmes, E. C., G. Pesole, and C. Saccone. 1989. Stochastic models of molecular evolution and the estimation of phylogeny and rates of nucleotide substitution in the hominoid primates. *Journal of Human Evolution* 18, 775-794.

Horton, D. R. 1984. Red kangaroos: Last of the Australian megafauna. In *Quaternary Extinctions*, ed. P. S. Martin and R.G. Klein, 639-680. Tucson: University of Arizona Press.

James, S. R. 1989. Hominid use of fire in the Lower and Middle Pleistocene: A review of the evidence. *Current Anthropology* 30, 1-26.

Johanson, D. C. 1989. The current status of Australopithecus. In *Hominidae: Proceedings of the 2nd International Congress of Human Paleontology*, ed. G. Giacobini, 77-96. Milan: Editoriale Jaca Book.

Jones, R. 1989. East of Wallace's line: Issues and problems in the colonisation of the Australian Continent. In *The Human Revolution: Behavioral and Biological Perspectives on the Origins of Modern Humans*, ed. P. Mellars and C. Stringer, 743-782. Edinburgh: Edinburgh University Press.

Kahlke, H. D. 1975. The macro-faunas of continental Europe during the Middle Pleistocene: Stratigraphic sequence and problems of intercorrelation. In *After the Australopithecines*, ed. K.W. Butzer and G.L. Isaac, 309-374. The Hague: Mouton Publishers.

Klein, R. G. 1984. The large mammals of southern Africa: Late Pliocene to Recent. In *Southern African Prehistory and Paleoenvironments*, ed. R.G. Klein, 107-146. Rotterdam: A.A. Balkema.

———. 1988. The causes of "robust" australopithecine extinction. In *Evolutionary History of the "Robust" Australopithecines*, ed. F. E. Grine, 499-505. New York: Aldine de Gruyter.

———. 1989a. *The Human Career: Human Biological and Cultural Origins.* Chicago: University of Chicago Press.

———. 1989b. Biological and behavioral perspectives on modern human origins in southern Africa. In *The Human Revolution: Behavioral and Biological Perspectives on the Origins of Modern Humans,* ed. P. Mellars and C. Stringer, 529-546. Edinburgh: Edinburgh University Press.

Kurtén, B. 1968. *Pleistocene Mammals of Europe.* London: Weidenfeld and Nicolson.

Kurtén, B., and E. Anderson. 1980. *Pleistocene Mammals of North America.* New York: Columbia University Press.

Lundelius, E. L. 1988. What happened to the mammoth? The climatic model. In *Americans Before Columbus,* ed. R.C. Carlisle 75-82. Ethnology Monographs 12, Pittsburgh: Department of Anthropology, University of Pittsburgh.

———. 1989. The implications of disharmonious assemblages for Pleistocene extinctions. *Journal of Archaeological Science* 16, 407-417.

Lynch, T. F. 1990. Glacial-age man in South America: A critical review. *American Antiquity* 55, 12-36.

Maglio, V. J. 1975. Pleistocene faunal evolution in Africa and Eurasia. In *After the Australopithecines,* ed. K.W. Butzer and G.L. Isaac, 419-476. The Hague: Mouton.

Marean, C. W., and D. Gifford-Gonzalez. 1991. Late Quaternary extinct ungulates of East Africa and paleoenvironmental implications. *Nature* 350, 418-420.

Martin, P. S. 1967. Prehistoric overkill. In *Pleistocene Extinctions,* ed. P. S. Martin and H. E. Wright, 75-120. New Haven and London: Yale University Press.

———. 1984. Prehistoric overkill: The global model. In *Quaternary Extinctions,* ed. P.S. Martin and R.G. Klein, 354-403. Tucson: University of Arizona Press.

———. 1990. Who or what destroyed our mammoths? In *Megafauna and Man: Discovery of America's Heartland,* ed. L.D. Agenbroad, J.I. Mead, and L.W. Nelson, 108-117. The Mammoth Site of Hot Springs Scientific Papers No. 1.

Martin, P. S., and R. G. Klein, eds. 1984. *Quaternary Extinctions: A Prehistoric Revolution.* Tucson: University of Arizona Press.

Martin, P. S., and H. E. Wright, eds. 1967. *Pleistocene Extinctions: The Search for a Cause.* New Haven and London: Yale University Press.

Mellars, P. 1989. Major issues in the emergence of modern humans. *Current Anthropology* 30, 349-385.

Morlan, R. E. 1988. Pre-Clovis people: Early discoveries of America? In *Americans Before Columbus,* ed. R. C. Carlisle, 31-43. Ethnology Monographs 12, Pittsburgh: Department of Anthropology, University of Pittsburgh.

Murray, P. 1984. Extinctions downunder: A bestiary of extinct Australian Late Pleistocene monotremes and marsupials. In *Quaternary Extinctions*, ed. P.S. Martin and R.G. Klein, 600-628. Tucson: University of Arizona Press.

Oates, J. 1980. The emergence of cities in the Near East. In *The Cambridge Encyclopaedia of Archaeology*, ed. A. Sherratt, 112-119. Cambridge: Cambridge University Press.

Potts, R. 1988. *Early Hominid Activities at Olduvai Gorge*. New York: Aldine de Gruyter.

Prentice, M. L., and G. H. Denton. 1988. The deep-sea oxygen isotope record, the global ice sheet system and hominid evolution. In *Evolutionary History of the "Robust" Australopithecines*, ed. F. E. Grine, 383-403. New York: Aldine de Gruyter.

Sheratt, A. 1980. The beginnings of agriculture in the Near East and Europe. In *The Cambridge Encyclopaedia of Archaeology*, ed. A. Sherratt, 102-111. Cambridge: Cambridge University Press.

Stringer, C. B. 1989. Documenting the origin of modern humans. In *The Emergence of Modern Humans*, ed. E. Trinkaus, 67-96. Cambridge: University of Cambridge Press.

_____. 1990. The emergence of modern humans. *Scientific American* 262(12), 98-104.

Stuart, A. J. 1982. *Pleistocene Vertebrates in the British Isles*. London and New York: Longman.

_____. 1983. Pleistocene bone caves in Britain and Ireland. *Studies in Speleology* 4, 9-36.

Susman, R. L., J. T. Stern, and W. L. Jungers. 1984. Arboreality and bipedality in the Hadar hominids. *Folia Primatologia* 43, 113-156.

Sutcliffe, A. J. 1985. *On the Track of Ice Age Mammals*. Cambridge, Mass.: Harvard University Press.

Tchernov, E. 1984. Faunal turnover and extinction rate in the Levant. In *Pleistocene Extinctions*, ed. P.S. Martin and H.E. Wright, 528-552. New Haven and London: Yale University Press.

_____. 1987. The age of the 'Ubeidiya Formation,' an early Pleistocene hominid site in the Jordan Valley, Israel. *Israel Journal of Earth Sciences* 36, 3-30.

_____. 1988. The biogeographical history of the southern Levant. In *The Zoogeography of Israel*, ed. Y. Yom-Tov and E. Tchernov, 159-250. Dordrecht: Dr. W. Junk.

Trinkaus, E. 1989. The Upper Pleistocene transition. In *The Emergence of Modern Humans*, ed. E. Trinkaus, 42-66. Cambridge: University of Cambridge Press.

van Donk, J. 1976. An ^{18}O record of the Atlantic ocean for the entire Pleistocene. *Geological Society of America Memoir* 145, 147-164.

White, T. D. 1984. Pliocene hominids from the Middle Awash, Ethiopia. *Courier Forschungsinstitut Senckenberg* 69, 57-68.

Williams, D. F., R. C. Thunell, E. Tappa, D. Rio, and I. Raffi. 1988. Chronology of the Pleistocene oxygen isotope record: 0-1.88 m.y. B.P. *Paleogeography, Paleoclimatology, and Paleoecology* 64, 221-20.

2

The Impact of Food Production: Short-Term Strategies and Long-Term Consequences

Charles L. Redman

Many of the rich and productive early civilizations of the world arose in what today are relatively uninhabited and degraded landscapes. This is particularly true of the arid and semi-arid regions of the Mediterranean, Mesopotamia, the American Southwest, central Asia, and the Indus River Valley. Why were complex civilizations once possible where today they are not?

There are times in the long history of the human career when seemingly intelligent, productive enterprises--efforts that improve human life in the short term--have negative, even destructive, long-term effects. One of the most important transformations in human history, the shift from nomadic hunting and gathering to settled agriculture and village life, is an important example of the difference between short-term strategies and long-term consequences. Peoples of ancient Greece, Mesopotamia, and the American Southwest, for example, engaged in various activities-- clearing land, planting seeds, irrigating crops, harvesting and storing those crops, and organizing themselves socially and politically to make those activities possible--that enabled more people to live better lives than their ancestors had. For a time. Then the long-term consequences of these seemingly wise endeavors emerged in the form of soil erosion, saliniza- tion, vulnerability to hazards such as floods and droughts, and political instability. And where large-scale, sophisticated civilizations once flourished, today we see aridity and emptiness.

This chapter is about the long-term consequences of short-term agricultural strategies. It begins with a discussion of some of the characteristics of early agriculture and its implications for prehistoric people who practiced it. Then three case studies are presented. The stories of now-gone civilizations in ancient Greece, the American Southwest, and Mesopotamia illustrate some of the tradeoffs involved in the transition to an agricultural economy. They also reveal some of the positive and negative relationships between people and their local environment that derive from a reliance on agriculture. Of particular interest are the decisions made to satisfy short-term objectives that in fact have unexpected consequences for long-term survival, of peoples and of civilizations.

This analysis focuses not on backward or even average agricultural societies, but on people responsible for some of the greatest cultural achievements of the past. The same people who brilliantly invented technologies, methods of subsistence, culture, and arts, appear thoughtless in the treatment of the environment. This study is not about poor, uneducated, culturally backward people, but those at the leading edge of invention and creativity.

It is not argued that the introduction of agriculture and the ensuing early civilizations were unfortunate developments. Quite to the contrary, the shift to agriculture may be the single greatest transformation in the human career. It has provided us with wondrous things and allowed us to reach our potential in many domains. Without food production, most of what we know as everyday life would not exist, and, for that matter, most of us would not exist. Yet, it is also true that the evolution of a food producing economy has taken a serious toll on people and on the world around us. Hence, one is neither a harbinger of doom nor a herald of joy when one tells the story of the development of agriculture. But there are certain implications of these experiences of the past, implications that help explain the current conditions of the world and provide lessons for the future.

Features of an Agricultural Economy

An agricultural or food producing economy can be defined as one in which there is a primary reliance for food on domesticated plants, animals, or both. Food production is comprised of four sets of activities, all of which must be present for a society to be fully agricultural. These four activities are plant cultivation or animal husbandry, harvesting, storage, and control of propagation.

In plant cultivation, natural vegetation is usually suppressed or removed, the biology of the topsoil is changed by hoeing or plowing, water is drained off or supplied, and weeds or predatory animals are controlled. Harvesting of course involves the gathering of the edible portion of the plant, perhaps for immediate consumption but sometimes for storage. It should be noted, however, that many non-agricultural peoples harvested plants that could be domesticated, or even cultivated these food resources before harvesting. It is widely agreed, however, that if these people did not control the propagation of these resources, then they were not fully agriculturalists. As long as a plant or animal maintains its capability for natural dispersal, it is likely that it could return to its natural state if the people manipulating it abandoned the locality. Hence, even intensive harvesters can have only a relatively impermanent impact on the environment. Yet once propagation is controlled, either through animal breeding or by selecting plants that require artificial dispersal, true agriculture is being practiced. Also, once propagation is controlled, the food resource and the environment are both profoundly affected.

The earliest villages to practice all four of these activities developed independently in various regions of the world between six and ten thousand years ago. I will describe some of the characteristics of such villages in the ancient Near East, the region I know best (Redman 1978). The typical early agricultural village in the Near East was located at an intermediate elevation in the uplands surrounding the great river valleys in areas where wild plants and animals existed that could be domesticated. Some villages also existed at lower and higher elevations. These were nucleated settlements of 50 to 100 people that under some circumstances grew to be much larger. The inhabitants built substantial multi-roomed, rectilinear homes. They also developed a diverse inventory of tools and containers, most notably grinding stones for food processing and ceramic vessels for storage. Even in these earliest villages we find evidence of increasingly specialized tools and facilities, the participation in long-distance exchange networks, and various activities that are best described as religious. These were truly innovative, gifted people who were taking part in one of the greatest experiments of all history.

These people affected the environment, but their impact remained minor. This was true most importantly because villages were few and widely scattered in localities favorable for rainfed agriculture. In addition, some strategies pursued by these early villagers helped preserve their local environment. At least in the Near East, we find villagers relying on a mixture of plant and animal resources. It is not certain, but we expect that their fields were not devoted to single crops, but were probably mixtures of both cereals and legumes that helped to maintain soil fertility. The decision to maintain a diversity of plant and animal resources, even in the smallest productive units, was probably a response to the high risk of relying on any single resource--any failure meant total failure. Diversifying minimized the chance of serious food shortages, and at the same time provided a balanced diet.

These Near East agricultural villages proved to be quite successful, enduring and being adopted throughout the Old World. Three advances were behind the success of this way of life. First were the physiological improvements in the domesticates. These include larger seeds, easier to process plants, and secondary products from animals, such as milk or wool. Second, the technology of food production, processing, and storage improved significantly with time. And third, it is expected that changes in the organization of human communities enhanced the effectiveness of the new productive economy. Whatever the exact reasons, village farming caught on and became the dominant settlement form from about six thousand years ago until the beginning of the twentieth century.

With the success of farming came serious disruptions in the local ecological balance, disruptions made irreversible by the nature of agriculture. As human populations grew, predators were killed, competitors were eliminated, land was cleared, and water was redistributed. All this occurred on an ever expanding scale and in such a way that reversion to a hunting and gathering way of life was made difficult and then impossible.

Implications of Agriculture

The successful introduction of agricultural economies had several significant implications. First, population growth was facilitated by the larger and more reliable food supplies. Not only was more food available more dependably, but a greater variety of food was available. More reliable sources of carbohydrates and a generally more balanced diet would lead to the earlier onset of menarche and a longer fertility period by raising the level of body fat, making higher birth rates possible. It is also possible that the new, more sedentary, lifestyle would encourage closer spacing of births. Better nutrition and a less nomadic life might also allow greater productive longevity of adults.

But while human populations were expanding, it is likely that animal populations were declining. The natural fauna of a newly farmed region would be seriously impacted through landscape transformation, capture, or killing. Declines would be most apparent among large herbivores who were our competitors and carnivores who preyed on our herds. The conflict between the wild fauna of a region and encroaching village agriculturalists that can be seen in many parts of Africa today may reflect processes at work elsewhere thousands of years ago.

The natural landscape and its flora were also seriously impacted by farming and herding. Local flora were replaced, some habitats such as swamps and forests were destroyed, some resources such as wood or certain stones were removed, and some agricultural practices led to serious soil erosion.

At the earliest stages of the agricultural revolution, the impact of the newly developed food producing economies did not have wide-spread or long-lasting negative impacts on the environment. This is largely because the population density of early agriculturalists was still quite low. Furthermore, their activities were small in scale and production was relatively modest. These early agriculturalists are unlikely to have recognized that they were taking part in the greatest revolution in human history. Both the agricultural improvements and the beginnings of environmental damage were the result of a set of small decisions made by unrelated groups to maintain or marginally improve their subsistence base with what they understood of the potential resources.

These new economies, however, contained the seeds of much greater environmental impact. The activities of the new agriculturalists not only served to maintain their new way of life, but also to maximize the short-term returns on their efforts. These are not illogical or destructive intentions. But these people in the past, as people today, appear to have made decisions based on a very short time horizon, without considering the long-term consequences of their actions. Three sets of activities, described below, doubtless improved food availability to those who initiated the processes. But viewed from the long-term perspective of the archeologist, these activities not only had certain social consequences that many would consider negative; in most cases, the activities led to the ultimate destruction of the society's subsistence base.

First, beyond the simple growth in population, there was a relocation of this increasing population into dense aggregations. Eventually, with the increasing importance of agriculture, there was a systematic shift in the distribution of settlements toward arable land and ultimately toward irrigable land. As part of this process there was a long-term growth in nucleated settlements, with some growing in density to urban proportions, exacerbating the impact on the local environment.

Perhaps equal to the immediate changes in the local ecology have been certain human organizational changes prompted by the new agricultural lifestyle. It is thought that family structure and familial relations must have changed dramatically with sedentism and the possibility of children and elderly working with herds and in the fields. With the building of permanent settlements, with food processing and storage facilities, and with the stored food itself there emerged an increased need for defense from raiders and an increasing temptation to raid others. This accumulation of physical goods and wealth also must have encouraged the idea of private property and led to differential access to these productive resources, creating the basis for class society. The sedentary lifestyle and the physical segmentation that came with the built environment of these villages and early cities may also have led to changing gender roles and possibly to restrictions in the activities of women and devaluation of their status.

Equal in importance with the reorganization of family and social life must have been agriculture's implications for productive efficiency and political organization. Archeological evidence makes it clear that even the earliest villagers were producing specialized tools and facilities to improve their efficiency. It must have been clear to them, as it is to us today, that specialization can lead to increased production. Yet, in the earliest villages we see that extra effort is expended to maintain a diversity of resources and productive activities. The flexibility that derives from diversity is an essential strength of most small-scale societies.

However, flexibility and specialization within the same unit are in apparent contradiction. The resolution that occurred in most societies is that flexibility and specialization are enacted at different levels. Specialization in productive activities takes place in local productive units. Flexibility of the system is maintained at a larger scale by integrating many productive units into, for example, a market-driven or administered economy. This conjunction requires that the component productive units become interdependent, that is, each unit relies on other units for some aspects of its subsistence and other necessities. The ancient civilizations, as is the modern world, were comprised of strongly interdependent productive units. Such arrangements are effective as long as interdependence works to enhance the long-term survival of its constituents and the maintenance of its environment. But interdependent systems can also be fragile systems vulnerable to external and internal pressures.

A brief summary of three case studies follows that illustrate some choices our ancestors made and how those choices affected their environment.

Ancient Greece

Ancient Greece is an area of considerable interest to any student of the past because of the greatness of its peoples and the strong influence they had on the formation of western civilization. Anyone who has visited Greece knows that it is a rugged, semi-arid country with relatively small alluvial valleys and accompanying slopes that are suitable for agriculture. Studies by geologists and archeologists have determined that numerous episodes of severe soil erosion have occurred during historic and prehistoric times. The type of erosion involved, sheet erosion, would be catastrophic for agriculture. It is known that erosion occurs when decreasing rainfall, deleterious human activity, or both reduce plant cover on the uplands. Scholars have argued recently that human misuse has been the major cause of soil erosion in ancient Greece as well as in the entire Mediterranean area (van Andel et al. 1990). The kinds of human activity found to accelerate soil erosion include clearing of land, farming the land, deforestation for timber, grazing of land, and fires. Accelerated loss of soils in the uplands then leads to catastrophic sedimentation episodes in the valleys.

By studying depositional stratigraphy in various valleys, these episodes of sedimentation can be dated and correlated with changes in local population densities and assumed intensity of land-use. Van Andel et al. found that, starting in the middle Bronze Age, periods of population growth were followed by episodes of extensive stream deposition of well-sorted materials. These episodes were in turn associated with periods of sharp decreases in the numbers of habitation sites in the general locality. Although precise dating is difficult, the authors suggest that periods of stable soils and population growth might last from one to several hundred years, followed by rapid deterioration, and a long period of low population in the area, before a return to improved agrarian conditions.

Van Andel et al. also suggest that the ancient human response to soil conditions may have been similar to what has been observed in historic times. In the more recent period, when soil conditions have stabilized in an area, more people move in, usually extending the areas that are farmed. Settlers often build terraces in the uplands to increase the surface area that can be used for farming. As more people occupy the area, fallow periods are shortened. The soil eventually loses fertility. The human response then is to withdraw to the areas with best soil, increasing the intensity of use there and to lease the poorer upland fields for pasturage, leading to damage to the terrace systems. Increasing farming intensity and damaged terraces both stimulate further upland erosion, eventually causing gully formation, stripping the remaining soil from the upland, and causing catastrophic deposition in the lowlands.

Although this cycle of degradation may be facilitated by changes in rainfall and sea level, Van Andel et al. are convinced that human disturbance of the slope equilibrium is the major culprit exacerbating natural climatic variations and leading to agricultural catastrophes. They believe that while it was physically possible to maintain the equilibrium of soils on the slopes even in the face of human use, the costs in terraces and farming techniques was high, and results usually did not last long. The most convincing evidence that the human hand, rather than climate change, is most responsible for the erosion and deposition episodes is this: Van Andel et al. have identified such cycles in various localities in Greece and neighboring eastern Mediterranean countries; these cycles occur at different times in different areas, while the climate changes were relatively uniform throughout the area.

It seems quite likely, therefore, that intensive use of arid landscapes accompanying population growth gave rise in ancient Greece to soil erosion that eventually depleted the very fertility that made productive agriculture and population growth possible in the first instance. It appears that the beginnings of soil erosion led in turn to the population aggregating into ever-denser settlements and to intensification of farming practices on the remaining fertile land. These strategies were doubtless appropriate in the short-term, but in the long-term they resulted in even more serious erosion and more loss of soil fertility and agricultural potential. Today, where the astonishingly creative and innovative culture of ancient Greece flourished, we find eroded slopes and degraded vegetation.

American Southwest

The prehistoric cultures of the American Southwest were complex, and they occupied diverse environments. Like Greece, the Southwest is a semi-arid region with rugged uplands and alluvial valleys. The geographic scale, soils, available plants, and average rainfall are all quite different from Greece. But the puzzle of prehistoric people occupying and farming an area that appears inhospitable today is present, as is the question of whether the ultimate deterioration of these areas was primarily the result of climatic or human factors.

The subjects of the next case study are the Hohokam of what is now central Arizona and the Anasazi of Chaco Canyon, in what is now northwest New Mexico. These cultures are considered by many to have attained the highest level of sophistication and political organization of any prehistoric southwestern societies. Yet the core areas of both were almost completely abandoned long before non-Indians entered the area.

The Hohokam were very effective irrigation agriculturalists who colonized the desert alluvial valleys of the Salt and Gila Rivers of central Arizona and their tributaries. They lived in clusters of semi-subterranean pithouses spread along the banks of canals they had dug to water their fields. At their peak, between 900 and 1200 A.D., some of their communities included hundreds of houses and had platform mounds and ball courts at their center. These people dug and maintained over one hundred miles of canals in the Phoenix area alone. Given its heat and aridity, the area that became the Hohokam heartland was not attractive to large numbers of settlers until effective irrigation technologies existed. But with irrigation, farmers were extremely productive and generated a sufficient surplus to support long distance trade and the construction of centralized buildings.

At first it appears that the Hohokam flourished in their river valleys for almost a thousand years. But closer investigation has led many scholars to conclude that the success of Hohokam occupation varied greatly--three or four cycles of population growth were each followed by serious depopulation. Recent studies of the growth cycles of trees in the upland watersheds of these rivers reveal a high variability in precipitation (Nials et al. 1989). Variability in rainfall would lead to variation in streamflow and both drought and flooding in the Hohokam area. Nials et al. suggest that major floods could destroy irrigation facilities that only significant construction efforts could replace.

Nials et al. offer the following reconstruction of events, based on their analysis of the tree growth data. After a major flood in 899 A.D., three centuries of relatively even streamflow followed, a condition favorable for the growth of irrigation systems and the populations they could support. During this time there may have been some minor destruction of facilities, but only on a scale that could be repaired quickly by the relatively well organized society that occupied the lowlands. However, beginning in roughly 1200 A.D., two centuries of far more erratic streamflow followed, with periods of low water followed by one or more seasons with catastrophic floods. This climatic pattern would have been too much for the Hohokam to handle. They would increasingly not be able to reconstruct their irrigation works, and would be forced eventually to abandon their homelands.

This strongly climatic explanation may be sufficient to explain the decline of the Hohokam. But a more subtle analysis, offered here, views the Hohokam decline more as the disruption of a balance between destructive forces of nature and the organizational ability of a society to maintain its productive facilities. Instead of saying that climatic forces are of such a scale that they overwhelmed the Hohokam, why not ask whether the society's ability to react effectively had been seriously undermined so that climatic variations that had been reasonably well handled in the past now became more deleterious? Two issues are relevant. First, during the Hohokam heyday we know that they traded widely and may have established colonies in upland areas. Settlement in the upland watershed may have led to local soil erosion, increasing runoff and the potential for severe lowland flooding. The second factor is the degree of salinization and waterlogging that the lowland areas experienced as a result of the long-term use of irrigation without the advantages of modern techniques, such as drainage tiles, to combat these problems. While nature certainly had a hand in the downfall of the Hohokam, the Hohokam may well have weakened themselves and their environment, partly because of their own success.

The Anasazi Indians of Chaco Canyon also achieved a level of material success that is widely cited as the pinnacle of Southwestern society. The core of this society was located in northwestern New Mexico close to the geographic center of the lower elevations of the San Juan Basin. In Chaco Canyon, an arid valley watered only by an impermanent stream, thirteen "great houses" of up to a thousand rooms each were constructed using amazingly fine masonry; numerous smaller sites were built as well. The major sites were occupied only a short time however, perhaps as little as 100 years, in the eleventh century. During this time the Chaco Canyon population grew to large numbers. It also appears that these people controlled or at least had significant interactions with communities all across northwest New Mexico and even further afield. Prehistoric roads lead from Chaco Canyon in all directions, and magnificent trade goods can be found in the canyon sites.

This portion of the San Juan Basin has always been a marginal zone for agriculture, averaging under 10 inches of rainfall a year in modern times. And Chaco Canyon in particular is not the most attractive site in the basin, with alkaline soil, no perennial stream, and a relatively cold climate. Why and how, then, did a great society flourish there?

Clearly the prehistoric people of the area worked very hard at increasing their productivity and reducing the possibility of subsistence stress. Actions they took included major storage facilities, water control mechanisms for maximizing the use of available runoff, population dispersal in the canyon and outside, and increased exchange to bring goods into the canyon (Vivian 1991). But hard work alone may not be the distinguishing feature of the Chacoan success; what may explain that success is their ability to get other groups to do their bidding.

Scholars vary in their population estimates for Chaco Canyon, but generally it is believed that enough people lived there that large quantities of goods must have been brought into the canyon, including foodstuffs. Thus what we see is the development of an organizational structure in which the Chacoans exerted either religious or political control over other groups who helped to provide the subsistence resources necessary for many of those living in the great houses. But such a political organization was quite new and probably unstable. As one might predict, the people of Chaco Canyon soon lost their influence and ceased to benefit from a far-flung network of suppliers. The great houses were quickly abandoned, and the valley became almost completely unoccupied.

Mesopotamia

Mesopotamia, the land of the Tigris and Euphrates Rivers, is the scene of the earliest and in many ways the most seminal ancient civilization (Adams 1978). This area, which is roughly coterminous with modern-day Iraq, hosted a series of early states and empires whose monuments and cultural achievements are among the most impressive in the world. Yet the core area of this civilization has now returned to desert conditions and is abandoned or inhabited only by relatively impoverished people. Although in our prior two case studies some argue (against the position offered here) that climatic change played the major role in destabilizing the societies, in Mesopotamia it is widely agreed that acts of the human hand led to its undoing.

This analysis will deal with two empires widely separated in time; the Ur III Dynasty of 2150-2000 B.C. and the Sasanian Empire of the early Middle Ages (226-637 A.D.). The Ur III Dynasty was focused in the southern half of Mesopotamia, and it consisted of a group of semi-autonomous cities inhabited by several tens of thousands of people each with an associated hinterland. This was a sophisticated society with well developed writing, a system of laws, extensive trade networks, and a period of strong centralized political control. The economic system relied heavily on irrigation agriculture with vast field systems along the Euphrates River and canals leading from it. Winter-cultivated cereals were the main crop, although there were many secondary crops. Herding was also important with contemporary records indicating as many as two million sheep being kept.

The aspect of Ur III society emphasized here is the rapid rise in the centralized control of the political hierarchy and paradoxically how that contributed to an era of declining agricultural productivity and environmental damage. Centralized control of the once independent city states was a logical objective of the growing power of the Ur III rulers. Centralization gave them greater access to labor pools, military conscripts, trade goods, and agricultural produce. More telling from our perspective, centralized control maximized production of food and other goods. Some of this increased productivity was achieved through increased specialization of production, but the majority resulted from centralized management of the construction and maintenance of water works and of the allocation of water in the growing irrigation network that fed the Mesopotamian fields. Moreover, it was a logical decision for Ur III rulers to attempt to extend the land served by irrigation and to increase the capacity of the existing canal system so more water could be brought to the fields. But the same decisions that brought short-term increases in production, as evidenced in the high population density and great construction projects of the Ur III period, rapidly undermined the agrarian base and led to a long period of diminished productivity. The major villain was salinization of the soils.

Salinization is caused by an accumulation of salt in the soil near its surface. This salt is carried by river water from the sedimentary rocks in the mountains and deposited on the Mesopotamian fields during natural flooding or purposeful irrigation. In southern Mesopotamia the natural water table comes to within roughly six feet of the surface. Excessive irrigation brings the water table up to within 18 inches of the surface. When soil is then waterlogged, as during flood irrigation, salt is carried by capillary action to the surface, killing most plants (Gelburd 1985).

Written records of temple storehouses of the period allow scholars to reconstruct with some certainty the relative productivity of fields and the crops being planted. A long-term decrease in productivity occurred between 2400 and 1700 B.C. At the outset of this period wheat was an important crop, accounting for at least 16 percent of the cereals produced. But as salinization increased, people slowly shifted to the more salt-tolerant barley, so that by the end of the Ur III Dynasty wheat made up only 2 percent of the crops grown. By 1700 B.C. it appears that wheat was totally abandoned as a crop in southern Mesopotamia (Jacobsen 1982). The end of this decline in wheat production coincides with a long period without centralized political control. Many cities were abandoned or reduced to villages, and the emphasis in agriculture shifted. Where before maximizing surplus production for central rulers dominated, later the object became satisfying the needs of local populations in a more self-sufficient production mode.

A brief look at the Sasanian case history brings some of these issues into clearer focus. The Sasanians had assembled a truly great empire that spanned most of Iraq and much of Iran. They built large cities and a very strong central government. Even more than the people of Ur III, they had come to rely on cereal cultivation from irrigated fields and had built a massive system of canals and facilities to bring water to an ever-increasing area of land. Although much uncertainty is involved, Adams (1978, 1981) has estimated that three to four times as much land was farmed during the Sasanian period as in the Ur III epoch. But, as under the earlier empire, this attempt at maximization led to several dangerous problems. First, the amount of water being brought to the land increased the risk of salinization even though the Sasanians devoted major efforts to combatting salt buildup. Second, the enormous scale of the Sasanian irrigation systems required more comprehensive and more effective management to succeed. Third, by extending cultivation to more marginal lands, the Sasanians both increased the amount of water being used and eliminated fallow lands that had been used for herding. This last point is particularly important in that the keeping of herds had always been a method of buffering against bad crop years as well as diversifying the diet.

Although the Sasanian Empire was large and lasted for several centuries, even its great organization could not hold it together in the face of these undermining processes. It is believed that diminished productivity was already taking its toll on central control when the region was hit by the plague several times in the sixth and early seventh centuries. This sickness further debilitated the system so that when Moslem armies entered Sasanian territory, the empire was toppled in 637 A.D. with surprisingly little resistance. The excesses of their farming system had weakened the once-great Sasanians.

The main point of these Mesopotamian examples is that in these pre-industrial states, political stability and economic maximization were only achieved by weakening the capacity of the productive system to react to internal and external challenges. State ideologies assumed at that time, as do many today, that everyone's interests are served when those of the central rulers are served. Yet we all know that "the people" may not share rulers' objectives and that all elements of the population may not benefit equally. What is less obvious is that those objectives may not serve in the long-term interest of the central rulers either! At least in the case of two Mesopotamian empires, strategies promoting short-term and long-term success were antithetical.

It is not necessary to end these Middle Eastern examples--or this chapter--on a note of doom, because after each of the cycles of centralization and collapse, a new state emerged in the same general region. This important resilience is likely the result of two major factors. One is that irrigation agriculture in alluvial valleys has an important regenerative capacity. Although salinization is a long-term problem, it can eventually be reversed, and less serious depletion of soil fertility can be corrected. Second is the ability of human systems to survive periods of social collapse. At least in Mesopotamia, the long-term survival of the society was insured by the existence of communities that assumed a marginal role in the centralized states. These communities maintained diversity in their productive activities and remained at low population levels. We believe that these communities existed at the margins of the empire even during periods of strong central control. But during the interims of weak central control, much of the countryside and population seems to have devolved to this level of existence.

It is the resilience of this more flexible way of life, in the face of changing times and serious threats, that has led to the long-term survival of many societies. The people who persist have maintained a flexibility in their productive activities and have remained at a modest level of productivity. While they do not attain great demographic or material achievements, they survive the natural and cultural challenges of history. They have clearly developed a strategy more in tune with their environment and less destructive of their resources than that of the great civilizations.

But before we all imagine that we should abandon our cities and return to small-scale self-sufficient settlements, consider what the world would be like without the achievements and inventions of those who chose to maximize. Most of us are thankful for the contributions of the great civilizations and forgive them their destructive actions. The important question is not which path to follow, maximizing or diversifying, but how to inform short-term strategies with their long-term implications. The lessons of the past do not reveal how that is to be achieved, only that the consequences of not doing so can be severe.

References

Adams, R. M. 1978. Strategies of maximization, stability, and resilience in Mesopotamian society, settlement, and agriculture. *Proceedings of the American Philosophical Society 122*, 329-335.

———. 1981. *Heartland of Cities*. Chicago: University of Chicago Press.

Gelburd, D. 1985. Managing salinity: Lessons from the past. *Journal of Soil and Water Conservation* 40, 329-331.

Jacobsen, T. 1982. *Salinity and Irrigation Agriculture in Antiquity*. Malibu, Calif.: Undena Publishing.

Nials, F. L., D. A. Gregory, and D. A. Graybill. 1989. Salt river streamflow and Hohokam irrigation systems. In *The 1982-84 Excavations at Las Colinas: Environment and Subsistence*, ed. D. Graybill, et al. Archaeological Series 162, Volume 5. Tucson: Arizona State Museum.

Redman, C. L. 1978. *The Rise of Civilization: From Early Farmers to Urban Society in the Ancient Near East*. San Francisco: W. H. Freeman and Co.

Smith, Philip E. L. 1972. The consequences of food production. In *Addison-Wesley Module in Anthropology* 31, 1-38.

Van Andel, T. H., E. Zangger, and A. Demitrack. 1990. Land use and soil erosion in prehistoric and historical Greece. *Journal of Field Archaeology* 17, 379-398.

Vivian, R. G. 1991. Chacoan subsistence. In *Cultural Complexity in the Arid Southwest: The Hohokam and Chacoan Regional Systems*, ed. Patty Crown and James Judge. Albuquerque: University of New Mexico Press.

3

The Epidemiology of Civilization

Mark N. Cohen

The word "civilization" usually brings forth images of moral, spiritual, artistic, and intellectual achievement and progress. If we stop to consider the penalties, if any, of becoming civilized, the list might include stress, loss of individuality, and the possibility of tyrannical leaders. Not entirely positive aspects of civilization should also include its effects on patterns of health and disease, or epidemiology. Indeed, the disease environment of civilizations is a predictable and recurrent concomitant of civilized society and very much of our own creation. It is as artificial and as essential as the civilized thought, art, technology, and architecture in which we take pride. And like the other issues described in this volume, it is a negative consequence of activities widely thought constructive and positive.

To study the epidemiology of civilization requires first that we examine the nature of infectious disease and of parasite-host interactions and second that we examine the nature and structure of civilization.

Infectious diseases do not result from the mere presence of infectious microorganisms in an environment. Instead, it is the relationship between the behavior of those organisms and human behavior that causes disease. Pathogens--the viruses, bacteria, protozoans, worms, and other parasites that cause disease--are living organisms with their own adaptive needs, survival requirements dictated by the rules governing their life cycles. They compete and adapt for their own survival and reproduction just as larger organisms do. They are not only active but reactive agents whose success or failure, and sometimes behavior, change as human habits change.

In some instances, other animal species such as mosquitos, flies, snails, cows, and sheep are also involved in the disease environment, serving as alternate "hosts", "reservoirs", or "vectors" (alternate homes and transportation) for the microorganisms. In that case, the behavior of those animals and their interaction with human behavior also affects the occurrence of disease. A change in human behavior that affects the survival or habits of either a microorganism or its vector may therefore facilitate or retard the spread of a disease.

These interactions of pathogens, reservoirs, or vectors with people are "regular" in the sense that they follow natural rules; we can to some degree predict or reconstruct the disease burden of a human population by knowing its location and its habits. We can predict disease in this way just as we can reconstruct prehistoric environments or predict the gravity of distant planets by using "uniformitarian" principles and extrapolating from contemporary experience. Different types of human social organization, even of populations living in the same natural environment at the same time, predictably generate different disease problems.

It is important to emphasize, moreover, that human behaviors may have profound effects on parasite cycles and therefore on disease whether or not they are consciously "medical" or health-oriented in their intent. Through human history, particularly prior to the twentieth century, non-medical behaviors have been of far greater consequence than medicine itself in determining the occurrence of disease. (Some modern sources suggest that intentional medical intervention probably did not significantly affect most diseases until the last 150 years at most. See McKeown 1976; Dubos 1965.) And since changes in our behavior can have negative as well as positive health consequences, encouraging as well as discouraging parasites, it should not surprise us that the history of medicine and health has been far less consistently positive or "progressive" than we are usually taught.

Some of the human behaviors that affect health may be random cultural choices, matters of local style and habit which can be adjusted with relative ease. With which hand do we eat? How, and how often, do we bathe? How high do we build our houses off the ground? From what materials do we build the house? When do religious rituals occur in the yearly cycle? Ecologically-oriented anthropologists have suggested that through long-term, often unconscious, trial and error, such minor cultural habits may be modified so that they are "adaptive," that is, they help to minimize the risk of infection (Alland 1970). For example, ritual bathing or taboos about the proper functions of the right and left hands or about the eating of a particular animal may protect people against the spread of particular parasites. In West Bengal, for example, it has been found that strict adherence to rules of ritual bathing helped to reduce hookworm transmission (Kochar et al. 1976). Inappropriate habits also occur. While Kochar et al. found a variety of personal habits that helped prevent transmission of hookworm, some, for example picking a popular, and therefore heavily infected place for bathing or picking a time of day when the worms are active, increased an individual's risk of exposure. (For further examples of the negative effects of behaviors see sources in Inhorn and Brown 1990.)

Many human behaviors that have affected the history of human health are not as simple or mundane as defecation; some are complex behaviors such as irrigation, urbanization, and trade, which feed and defend growing populations and are inextricably linked with the civilizing process. These behaviors are not easily modified or eliminated even when they are found to threaten health. In these instances, "adaptive" behavior may involve a compromise between health and politics or even between competing health needs. Human history has involved a long series of compromises among health needs, political competition, and increasing population; we must explore these compromises if we are to understand the epidemiological dimensions of that history.

Civilization: Its Structure and Ills

Although considerable controversy rages about how civilization originated, a fairly good consensus exists about how it is defined. These definitions make its epidemiological importance easier to grasp (see for example Fried 1967; Service 1975).

To anthropologists, civilizations are defined and distinguished from other kinds of societies by their very large size (historically the major factor in their competitive success); their heterogeneity of population and of communities; their tendency to be organized around craft specialization as much as or more than by kinship or family ties; their stratification into markedly distinct social and economic classes; their relatively fixed boundaries; and their common tendency to be ruled by centralized governments that hold a monopoly on the right to use physical coercion to maintain order. Perhaps more important from a medical point of view, civilizations are also defined by the large scale transformation of the natural landscape undertaken to create them (see for example Redman, this volume); by the use of large scale trade to bind together and support the heterogeneous or specialized communities of which they are comprised;

and by the existence of some specialized communities--the larger cities--
that may reach unprecedented sizes and population densities (and may
become wholly dependent on trade both for food and for the raw materials
to support their craft specializations).

One key point to be emphasized is that this structure seems to appear
primarily as a function of political competition with neighboring groups.
States commonly form in the presence of other states or in the presence of
other political units that are themselves approaching statehood (Fried 1967;
Stevenson 1968; Carneiro 1970; Sanders and Webster 1978). State
formation seems in large part to be a kind of arms race; and, like any
form of arms race, it is probably not motivated by concerns for the health
and well being of its individual members. Indeed, a state may flourish
despite its citizens' health problems.

The effects of this structure on infectious disease can best be described
if we examine the effects of certain characteristics of civilizations. (We
consider these characteristics separately, although in fact combinations of
factors are involved in the etiology of most diseases.) We consider in turn
the effects of population density, human effects on vector and reservoir
species, sedentary habits, animal domestication, changing diets, hygiene,
and trade and transportation.

The Effects of Population Density

Perhaps the single most important principle of the epidemiology of
civilization is that the success of most pathogens depends on the number
and density of the host organism, that is, the victims. This is true of plant
and non-human animal pathogens, as well as those that attack people.
Increases in the density of the human population and in the size, density,
and clustering of human groups, although not uniform or universal, are
two of the most obvious trends in human history (Cohen 1989). Indeed,
they are widely recognized as essential ingredients of the emergence of
civilization (Fried 1967; Stevenson 1968; and Carneiro 1970). An
increase in the number of human hosts increases the probability that the
offspring of a pathogen can successfully infect a new host, thereby both
increasing the number of pathogenic organisms present and reducing the
probability that the pathogen will die out. From the point of view of the
human host, this means that as human population grows, a pathogen is
more likely to be present--and continuously so--and that an individual
human being is more likely to be infected. The individual is also likely to
be reinfected more often; is likely to be infected by larger doses, thus

accumulating a larger disease burden or "parasite load;" and is likely to be exposed to a larger variety of pathogens. Since the presence of one infection is likely to facilitate infection by a second organism, the disease effect of larger and denser host populations is self-compounding.

These effects can be observed readily in archeological populations and in ethnographic populations (as summarized in Cohen 1989). Among archeological populations, the frequency of infection almost universally increases with the size and aggregation of the human population (see studies in Cohen and Armelagos 1984). For example, tuberculosis, one of the major diseases most visible in prehistoric populations because it often attacks the skeleton in characteristic ways, occurs almost invariably in relatively recent, large, urbanized populations (see also Buikstra 1981). Staphylococcal and streptococcal infections on bone surfaces (periostitis) also increase quite regularly with population density in various parts of the world, as do deep bone infections (osteomyelitis) and infection with the treponema bacteria that cause yaws.

Epidemiologists speculate that the effect of host population density on disease is so powerful that only a limited number of infectious diseases could have attacked early human populations. Early humans lived in small groups that rarely communicated with other groups, and population densities were low. Several analysts suggest (Black 1975; Cockburn 1971; Hare 1967) that only chronic diseases, especially those transmitted by direct human contact, and diseases caused by organisms whose life cycle does not require human beings could have survived these circumstances.

The organisms causing chronic diseases, for example the herpes virus, Epstein-Barr virus, and the yaws bacteria, remain active and infective in the individual human body for very long periods of time (often the better part of the lifetime of the human host) and thereby retain their capacity to infect other people for years. The parasite can survive and ultimately disseminate its young despite the small number of potential new hosts available. Also, chronic diseases transmitted directly and reliably by intimate contact between individual hosts--again, such as herpes and yaws-- enjoy a very high probability of successful transmission from host to host, helping offset the small number of possible hosts.

Another characteristic of chronic disease is important for this discussion. These infections are often relatively mild and, although widespread in a group, cause relatively little serious illness. The mildness of the disease they cause is in fact the key to their ability to survive in small groups--organisms that kill their only hosts do not last long.

Small groups of people may also be infected by organisms whose life cycles normally do not involve people at all, so that they can survive long periods without any human intervention. Such organisms may "accidentally" attack the occasional human being who comes within range. Included in this category are soil-borne organisms including various pathogenic fungi such as blastomycosis or coccidioidomycosis. (The latter is an infection inhaled from soil dust of the American Southwest, and evidence for its existence is thought to be found in archeological samples. See Selby 1975a,b; Steinbock 1976; Ortner and Putschar 1981).

Also included in this category--and of greater importance--are "zoonoses" or diseases that normally infect and circulate among other animals. These diseases include tularemia, toxoplasmosis, rabies, trichinosis, zoonotic malaria, various forms of encephalitis, bubonic plague (normally a disease of wild rodents), and a wide variety of others (Hubbert et al. 1975).

In striking contrast to the chronic human diseases, the zoonotic diseases generally do not spread from an infected person to another person. Typically, only one or a few individuals are affected at any one time, and infection can be renewed only through a new contact with the appropriate animal or soil. Also in contrast to the chronic diseases, these diseases are often severe. The human population often has had relatively little prior experience with the disease through which to develop resistance or to learn appropriate behavior. Moreover, since the organism has not been under the pressure of natural selection for prolonged coexistence with people--a selection that would favor organisms that, while making their hosts sick, do not kill them--its attack may be relatively virulent. These diseases may have profound effects on small groups, even though a small number of people are likely sick at any given time. The diseases typically are lethal or incapacitating, and they tend to attack adult individuals who make important economic contributions to the group and who are less easily replaced than children (Goodman et al. 1975; Cohen 1989).

Work by epidemiologist Francis Black (1975; Black et al. 1974) and others studying antibody patterns of small isolated contemporary tribes has tended to confirm these predictions about disease patterns. His studies of blood antibodies (which provide a lifetime record of a portion of an individual's infectious disease experience) show that chronic diseases tend to be present universally in such groups. In contrast, zoonotic diseases display patterns suggesting random exposure, the cumulative risk being proportional to an individual's age.

A key point in the epidemiology of small groups is that epidemic diseases, which are both acute or lethal and widespread, do not occur, apparently because small groups cannot sustain them. In his study groups, Black found that antibodies for acute epidemic diseases occurred only sporadically and displayed age-related distributions, occurring in all individuals above a certain age but in no younger individuals. He concluded that these diseases were not present as a constant problem in small populations because the organism apparently could not survive in small groups; nor did these diseases appear to be a constant threat in the natural environment. Instead, brief, self-limiting epidemics would occur only when the group experienced outside human contact, which, Black argued, was probably with the encroachment of civilized groups representing reservoirs of disease.

The Effects of Human Activities on Vector and Reservoir Species

A second major principle in the evolution of disease is related to the impact people have on other animals that normally act as alternate hosts or as vectors for infectious agents. If human activities make the environment inhospitable for vector or reservoir species or reduce their geographical range, vector-borne diseases decline in importance. If we expand the natural environment available to vector species, increase their proximity to human activities, or otherwise improve their survival and reproduction, we increase the probability of infection.

Many types of mites and ticks that carry rocky mountain spotted fever, scrub and tick typhus, and a host of related diseases around the world (Burgdorfer 1975) are primarily inhabitants of unimproved wild environments that do not react positively to human alteration of the landscape. For the most part, our behavior has done little to increase the risk of these tick-borne diseases. The risk is primarily to hikers, hunters, and campers, people who temporarily adopt uncivilized lifestyles, or to those whose domestic pets bring ticks home. Lyme disease, which has spread rapidly in the United States in the last fifteen years, is an interesting exception to this generalization. However, its increase appears to result at least in part from the resurgence of the wild deer population, which provides a primary host for some implicated species of ticks.

In contrast, many species of mosquitos, which are among the most important carriers of malaria, yellow fever, and dengue fever, appear to be encouraged by human activities such as forest clearance, irrigation, creation of standing or stagnant water, and, in some cases, house construction (Desowitz 1980). At the turn of the century, epidemiologists working to eliminate yellow fever during the building of the Panama Canal realized that standing water resulting from human modification of the landscape--including features as trivial and ephemeral as flower pots on hospital grounds as well as other more essential features of water management or drainage--was a major source of mosquito breeding and therefore of infection (McCullough 1977). More recently, anthropologists have argued that malaria, particularly falciparum malaria, the most lethal variety, was almost certainly encouraged by the Neolithic revolution, the essential prelude to civilization in which human groups adopted more sedentary habits and undertook large-scale land clearance for farming (Livingstone 1958, 1964, 1984).

Standing water is also associated with schistosomiasis, which with malaria is probably the most widespread and economically important of infectious diseases around the world today. This disease is transmitted to human beings in standing or slow moving fresh water, where certain species of snails live. People are infected by standing or bathing in the water, and they in turn contaminate the water with the pathogen by defecation or urination. The snail is essential since the parasite must live in the snail before it can reinfect a human being. People have enormously extended the distribution of shallow standing water necessary for the survival of both snail and parasite, largely through the creation of irrigation ditches and paddy fields; and it is primarily agricultural workers who are affected by the illness (Nelson 1975).

The Effects of Permanent Human Settlements

The adoption of sedentary habits by human populations also increases the importance and particularly the intensity of infectious diseases. This is so partly because permanent settlements enable parasites or vectors to maintain continuous contact with human hosts and partly because people affect the behavior of reservoir and vector species. Sedentism may reduce the range of zoonotic diseases to which populations are exposed, but it increases the number of organisms that can complete their life cycles in proximity to human groups.

The effects of sedentism are particularly important for parasite or vector species that must--or at least usually--spend part of their life cycles outside human hosts. For example, fecal-oral diseases, which cause most diarrheas and which arise from oral contact with fecally contaminated water, food, dirty objects, or hands, are minimized among groups that abandon contaminated campsites at frequent intervals; but they increase in importance when groups opt for permanent settlement. Sedentism in combination with large populations and centralized water and sewage supply seem to be the essential conditions for the transmission of severe acute fecal-oral infections like cholera, which are essentially diseases of recent and modern urban populations.

As one indication of the connection of disease and sedentism, weanling diarrhea, the debilitating combination of fecal-oral infection and malnutrition afflicting toddlers, considered to be one of the major killers of small children in the modern Third World, is generally not found or at least not common among the few known modern mobile hunting groups (Jelliffe, et al. 1968; Cohen 1989). Moreover, it was apparently not as significant for prehistoric hunter-foragers in most parts of the world as for their more civilized successors. Enamel hypoplasia, permanent age-specific stress markers on tooth enamel that commonly form at or near the age of weaning and are associated with weanling diarrhea at least in some modern populations, are generally less common among mobile prehistoric hunter-foragers than succeeding, sedentary farming populations in prehistory (Cohen and Armelagos 1984; Cohen 1989). This may explain why infant and child mortality rates among the most primitive known groups are not particularly high by historic standards (Cohen 1989). And it may be one reason why infant and child mortality rates in prehistoric populations do not appear to be as high in proportion to adult mortality as is true of modern populations (Lovejoy 1977; Cohen 1989).

Hookworm (infection by the worms Necator and Ancylostoma, which spread their eggs in human feces and reinfect other human hosts through the skin) is even more clearly linked with sedentism. The worms must spend a portion of their lives in the soil before they are infective; mobile groups are more likely to leave them behind. In fact, contemporary mobile groups commonly have relatively light infections, and the skeletons of prehistoric mobile groups commonly display comparatively low rates of porotic hyperostosis, an indicator of hookworm (and several other) infections (Cohen 1989; Cohen and Armelagos 1984).

The spread of human malaria is similarly limited by group mobility, especially when groups are small. An individual mosquito must bite at least two people in succession and at an interval of several days to effect transmission, because the parasite goes through a developmental stage in the mosquito. But mosquitos fly neither far nor fast and are incapable of following a moving human population.

Fleas and their diseases are also discouraged by human mobility (or mobility of other animals) because, although the adult organisms are body parasites that travel with their hosts, the larval and pupal stages are not. In these stages they live and develop in bedding or furnishings, and the reproductive cycle tends to be broken when people abandon old campsites (Anderson 1975).

Another effect of sedentism results from the fact that sedentary people build more substantial houses and move more of their activities indoors than do nomads. Some microorganisms, particularly airborne ones, are transmitted more effectively indoors, where they are not killed by sunshine or washed away by air currents and where the effective human population density is relatively high. In the modern context this may account in part for the tendency of diseases such as influenza to spread more readily and be more common in winter (Mogabgab 1976).

Two further facets of sedentary life, the necessity to store food and the inevitability of accumulating garbage, also provoke a significant change in the list of potential vector and reservoir species that become regular companions of human populations. Rats, whose diseases pose only occasional risks to human groups in the wild, generate a significant risk when they become regular inhabitants of human settlements. Rats and the parasites that they carry in their hair and on their skins transmit a variety of little known infections including rat-bite fever, hemorrhagic fevers, and a tapeworm (hymenolepis) (Hubbert et al. 1975). They are also occasional vectors of a wide range of diseases, such as rabies, that we usually associate with other modes of transmission. The best known danger associated with rats, bubonic plague, is transformed from a sylvatic zoonosis with rare human victims to a major human "plague" primarily because large human populations and large associated deposits of stored food and garbage attract large, concentrated rat populations (Davis et al. 1975). Rats and their fleas also transmit one form of typhus and appear to be the original source of epidemic human typhus. Typhus became a disease of people and their body lice (killing the lice as well as the people) only under conditions of extreme crowding and filth (Burnet and White 1972).

The Effects of Animal Domestication

The domestication of animals appears to have been a major source of new human diseases. Domestic animals may simply bring zoonotic diseases with them when they enter human society, or, remaining semi-wild, they may act as a continuous bridge between wild and human communities. In this manner, domestic dogs can bring tick-borne diseases or rabies into the community (Sikes 1975), and cats can bring toxoplasmosis, a disease potentially threatening to the human fetus. (See McCullough and Remington 1975.)

Some parasites whose life cycles involve both human and animal hosts are particularly encouraged by the continuing proximity of people and specific domestic animals. Tapeworm life cycles commonly involve both a carnivore (or omnivore) and a herbivore. Carnivores pass the eggs to herbivores in their feces and are reinfected in turn by eating the flesh of the herbivore. The intimacy of the association between the two hosts and the population density of each affect the probability of successful transmission. For example, beef and pork tapeworms pass to people who eat those meats and in turn pass to cows and pigs eating food contaminated by human feces (Rausch 1975). Completing such a cycle successfully and repeatedly is improbable when small groups of people occasionally hunt mobile herds rarely exposed to human feces. The probability of success for the parasite increases dramatically, however, when large numbers of animals are kept permanently in intimate proximity to the people who eat them. Other diseases increase in importance with crowding and enclosure of herds, the movement of herds from field to field, and the movement of animals from one herd to another. Tetanus, the potentially fatal anaerobic bacteria found in "dirt," occurs in many animal and human intestines and is only a contaminant of soils after infected domestic animals have passed or grazed (Rosen 1975).

Even more important, we believe that domestic animals are the ultimate source of many modern human infections, including most of the major epidemic diseases. Like people, disease organisms have evolutionary roots. We trace those roots using essentially the same principles of structural similarity to establish family relationship that are used to trace our own ancestry. A number of organisms thought to be relatively new as human diseases display similarities to viruses and other parasites that infect domestic animals. For this reason, smallpox and some forms of tuberculosis are thought to have come from cattle (although tuberculosis may also have originated in the New World in the absence of domestic cattle. See Clark et al. 1987; Allison et al. 1973). Measles appears to come from dogs or cows; influenza from pigs, chickens, or ducks; and the common cold from horses (Fiennes 1978; Cockburn 1967, 1971; Hare 1967; Mims 1980). The human diseases are thought to be mutant forms of their animal counterparts. While it is possible that such a mutant might spread from a wild animal to a human hunter, the probability of transmission obviously increases enormously with sedentism, when large groups of people are in continuous contact with large groups of animals, exposed to the full range of animal excretions, tissues, and fluids on a regular basis.

The Effects of Changing Diet

A large number of contemporary studies have suggested that quality of diet interacts significantly with disease experience. It is true that a few studies suggest that poor diet may be protective, helping to prevent or ameliorate particular diseases. For example, a diet poor in protein, iron, or any of several vitamins is thought to protect individuals against malaria; famine has been observed to have malaria-suppressing effects (Beisel 1982; Scrimshaw, et al. 1968). However, the vast majority of interactions between infection and malnutrition appear to be synergistic, with poor nutrition exacerbating infections and infections depleting nutrition (see Scrimshaw et al. 1968; Taylor 1983; Beisal 1982).

In this context it is worth noting that the average quality of human nutrition probably declined as people became more civilized. Contemporary "primitive" hunter-foragers have relatively well balanced diets, rarely deficient in vitamins, protein, or minerals. They are frequently marginal and occasionally deficient in caloric intake by Western standards but generally not in comparison to Third World averages. (Third World farmers may produce calories more efficiently than hunter-foragers but appear to lose a larger fraction of their production to landlords and the apparatus of the state.) Moreover, studies of the efficiency of hunting and gathering techniques suggest that caloric returns for hunter-foragers would have been higher early in prehistory when large game animals were more readily available and in spite of the fact that weapons gradually improved through time (Cohen 1989).

Prehistoric skeletons tell the same story. Early hunter-foragers seem to have been relatively large individuals--the prehistoric trend in human stature is generally downward, not upward. Moreover, other signs of nutritional stress such as porotic hyperostosis (anemia), premature osteoporosis, and growth retardation in children seem to become more frequent in later populations (Cohen 1989; Cohen and Armelagos 1984). Declining nutrition is therefore likely to have contributed to the increasing health problems of civilizations and to the incidence of parasitic infection.

The Paradoxical Effects of Hygiene

One of the results of the rise of civilization and the increasing burden of infectious disease that accompanies it is an increasing awareness of contagion and of hygienic principles. Hygiene emerges not simply because people are more "civilized" and knowledgeable but because the epidemic diseases of civilization are more obviously "caught" from other people than either the chronic diseases or the zoonoses of smaller groups.

Hygienic rules for the most part have positive effects in slowing the transmission of infectious diseases, but not always. For example, one theory of the origin of syphilis suggests that it developed or evolved from its close relative yaws, a skin disease transmitted by direct skin-to-skin contact and whose causative organism is indistinguishable from that of syphilis. When clothing and hygiene interfered with normal, universal childhood transmission, the spirochete causing yaws was forced to pass between individuals later in life through venereal contact (Hudson 1965). By this argument, syphilis is primarily a disease of colder climates but also of the kind of rules of social distance that accompany civilized lifestyles.

Polio, a disease brought to prominence in this century by the affliction of Franklin Roosevelt--and which was the great disease-scare of childhood in my generation until a vaccine was developed--has a related history. The organism that causes the disease is normally a fairly common inhabitant of the intestines of children, causing mild diarrhea. Antibodies to the virus, indicating exposure, are common in blood sera of children in tropical countries. But as hygiene improves, individuals on the average are infected progressively later in life, and it is this late age of infection that is likely to result in paralysis (Burnet and White 1972). Paralytic polio is primarily a disease of wealthier individuals, like Roosevelt, in societies that maximize the hygienic isolation of their young. Similarly, Epstein-Barr virus is a very regular companion of human groups around the world; but it appears to cause infectious mononucleosis primarily in cultures in which hygiene delays infection until puberty (Miller 1976). (On the other hand, early exposure to E-B virus may be associated with Burkitts lymphoma, helping to explain why this cancer is primarily a cancer of Third World countries. See Marx 1986.)

The Effects of Trade and Transportation

Human agencies not only enable new diseases to survive and flourish, they also play a major role in transporting them from region to region. Measles, mumps, smallpox, typhus, possibly tuberculosis, and plague are thought to have spread from Europe to the New World through the voyages of Columbus and later explorers (Crosby 1972; Dobyns 1983; Ramenofsky 1987). Malaria, particularly the more virulent forms, is thought to have spread from Africa to the New World through the movement of explorers and slave ships in the 16th and 17th centuries

(Ramenofsky 1987). Yellow fever may also have come to the New World from Africa, although a stronger case has been made for its ancient presence in the New World (Ramenofsky 1987). But wherever the origin, yellow fever and dengue fever are both thought to have been disseminated widely around the world in sailing ships.

Baker and Armelagos (1988) have recently made a strong case for the argument that syphilis spread from the New World to Europe at the time of Columbus. The argument depends on the fact that the bacterium that causes syphilis and yaws leaves characteristic lesions on skeletons and is therefore recognizable in archeological sites. Such lesions occur commonly among American Indian groups before Columbus. However, they are unknown in European, Asian or African skeletal collections dating from before this time. These data tend to overrule vague pre-Columbian literary references in the Old World, including passages in the Bible that are thought to refer to syphilis.

Within Europe, the transmission of bubonic plague appears to have been intimately associated with trade routes as well as with urbanism. Biraben (1968) has demonstrated a strong correlation between community size, position in trade networks, and mortality from plague among French cities of the early 18th century.

More recently, cholera is thought to have been spread around the world through the movement of Moslem religious pilgrims and of British military forces (McNeill 1980). The massing of huge numbers of troops for World War I and their subsequent dispersal seems to have played a key role in the great influenza pandemic of 1918. Labor migration in Africa still appears to be instrumental in the movement of strains of malaria; and schistosomiasis commonly moves from one irrigation system to another through the migration of workers. And AIDS, of course, may spread from person to person by sexual activity, transfusions, and needles, but it is spread from city to city by rapid transit and from continent to continent by airplane.

Critical Mass, Epidemic Diseases, and Competitive Success of Civilizations

For the first time in human history, in the last few hundred or few thousand years, civilizations seem to have provided the critical mass of population necessary to enable some acute viral infections to survive and spread continuously within the human community. Large urban concentrations of people who trade widely with other populations and engage in activities such as the domestication of animals are thus seen as behaving in ways that increase the incidence of disease and account for much of the history of infectious disease. In turn, this disease environment accounts for important features of the political history of the last several centuries.

Viral diseases like measles (and to varying degrees smallpox, mumps, German measles, influenza and even the common cold) run a singular course in the body. Their attack is of short duration because the disease either kills the host or provokes such a powerful, permanent immune reaction that it is defeated by the host's body. In either case, dead or permanently immune, a particular host is of no further use to the pathogen after a few weeks. The disease must spread continuously and rapidly from person to person to survive, in much the same way that a fire constantly needs new fuel. It has been estimated that a population of several hundred thousand individuals in reasonably close contact with one another would be necessary to maintain sufficient "fuel" for these diseases; and it is generally assumed that such diseases could not survive and spread in a world peopled by a relatively small number of widely scattered individuals or groups (Black 1975; Bartlett 1966). A network of very large populations connected by frequently used transport would be necessary to keep such diseases permanently housed in the population.

Work with antibodies (Black 1975, Black, et al. 1974) further supports the idea that in small, relatively isolated groups these diseases (caught presumably by contact with larger, more "civilized populations) do in fact "burn out" after a brief period when there are no remaining susceptible individuals or when new potential victims (immigrants and new babies) are too few to feed the steady demand.

It is also possible that the viruses producing the major epidemic diseases through human history have appeared repeatedly from animal sources, since the mutations that produce them, like most mutations, are likely to be repeated events. In early society the mutation would likely have only a brief existence, dying out after afflicting a few individuals. But in the civilized world such disease organisms were viable on a continuing basis. Though subject to repetition, mutations are relatively rare, and the appearance of epidemic diseases would have lagged sometime after the emergence of critical population thresholds. Therefore, it is quite possible that areas of the world like Mesoamerica, where the history of dense population was relatively short, might have remained free of epidemic diseases until infected by outside contact (Ramenofsky 1987).

The political impacts of these characteristics of epidemic diseases result from the different pattern of effects that these diseases have on "primitive" and "civilized" societies. Within civilizations, the major epidemic diseases quickly become "diseases of childhood." They circulate at rapid intervals and only children born since the last epidemic lack immunity. Protected by their parents and siblings and the ongoing, intact structure of a society run by immune adults, exposed children have a good chance of surviving. (In the early stages of civilization and as late as the nineteenth century, the pattern of epidemic disease probably generated more child mortality or a higher ratio of child to adult mortality than had previously been the case.)

In communities without prior exposure, however, these diseases attack individuals of all ages. Not only is the attack severe because of complete lack of immunity, but the elimination of parental care and social supports adds to the crises. For example, a virgin soil epidemic of measles was witnessed among the Yanomamo of South America in the late 1960's (Neel, et al. 1970). Large numbers of people died not just of measles but of complications from filth, malnutrition, and other secondary effects. The pattern of attack and the consequent accompanying suppression of behavioral response by the groups may account for historically high death rates among American Indians from these diseases, even in the absence of any apparent genetically-based deficiencies in immunity. In other words, group size and disease history, not genetically mediated immunity, determined the pattern of sickness (Neel 1977).

The differential impact of epidemic diseases rapidly became a major political factor. Epidemic disease was not only a defining feature of civilization, it became a major weapon in its arsenal and a factor in its adaptive success. To a large extent the spread of civilization at the expense of other social forms in the last few thousand years--and certainly the ability of small groups of sixteenth-century Europeans armed with very primitive firearms to "defeat" vastly superior forces of Amerindians-- results largely from this advantage (McNeill 1976, 1980).

An Alternate View of Human History

The logic of microorganisms outlined in the preceding section, combined with the evidence of skeletons and ethnographic observations of small scale societies discussed in earlier sections, all suggest that disease has increased in significance as a contributor to both morbidity and mortality as populations have become more civilized. Data from the same sources also suggest that the quality, quantity, and reliability of human diets declined in the course of the same developmental sequence, leading to further health problems.

Even though the average life expectancy of adults (those who have already reached age 15) appears to increase "progressively" through prehistoric and historic time in the Old World (at least after the Neolithic Period and with major exceptions in major cities), there is in fact no evidence from any source to suggest that infant and child survivorship (the percentage of live births surviving to age 15) improved until the latter half of the nineteenth century or later. The recorded percent of infants and children lost in areas of Europe in the eighteenth and early nineteenth century, well in excess of 50 percent, are generally as high as or higher than numbers reported from archeological or ethnographic samples of uncivilized societies. Life expectancy at birth in Europe in the eighteenth century--averaging between 20 and 30 years--is not substantially different from the best available archeological or ethnographic estimates in the simplest and earliest hunter-forager populations (summarized in Cohen 1989).

The spread of civilization, therefore, has a different basis than we generally believe. The success of early civilized systems results in part from the paradoxical fact that they were reservoirs of new and deadly diseases to which the less civilized were not exposed. Thus the success of the political systems of civilizations should not necessarily be confused with the success of the individuals who participate in those political systems. Civilization may well have spread competitively despite making relatively heavy demands on the health and welfare of its bearers.

References

Alland, A. 1970. *Adaptation in Cultural Evolution.* New York: Columbia University Press.

Allison, M. J., et al. 1973. Documentation of a case of tuberculosis in preColumbian America. *American Review of Respiratory Diseases* 107, 985-91.

Anderson, G. S. 1975. Schistosomiasis. In *Diseases Transmitted from Animals to Man*, ed. W. T. Hubbert, et al. Springfield, Ill.: C. C Thomas.

Baker, B. J. and G. J. Armelagos. 1988. The origin and antiquity of syphilis. *Current Anthropology* 29, 703-21.

Bartlett, M. S. 1966. The critical community size for measles in the United States. *JRSSA* 123, 37-44.

Beisel, W. R. 1982. Synergisms and antagonisms of parasitic diseases and malnutrition. *RID* 4, 746-55.

Biraben, J. N. 1968. Certain demographic characteristics of the plague epidemic in France, 1720-22. *Daedalus* 97, 536-45.

Black, F. L. 1975. Infectious diseases in primitive societies. *Science* 187, 515-18.

Black, F. L., et al. 1974. Evidence for persistence of infectious agents in isolated human populations. *American Journal of Epidemiology* 100, 230-50.

Buikstra, J. E., ed. 1981. *Prehistoric Tuberculosis in the Americas*. Evanston: Northwestern University Archaeological Program.

Burgdorfer, W. 1975. Introduction to rickettsioses. In *Diseases Transmitted from Animals to Man*, ed. W. T. Hubbert, et al. Springfield, Ill.: C. C. Thomas.

Burnet, Sir McF., and D. O. White. 1972. *Natural History of Infectious Diseases*. 4th ed. Cambridge: Cambridge University Press.

Carneiro, R. 1970. A theory of the origin of the state. *Science* 169, 733-38.

Clark G. A., et al. 1987. The evolution of mycobacterial disease in human populations. *Current Anthropology* 28, 45-60.

Cockburn, T. A. A. 1967. *Infectious Diseases: Their Evolution and Eradication*. Springfield, Ill.: C. C. Thomas.

———. 1971. Infections diseases in ancient populations. *Current Anthropology* 12, 45-62.

Cohen, M. N. 1989. *Health and the Rise of Civilization*. New Haven: Yale University Press.

Cohen, M. N., and G. J. Armelagos. 1984. *Paleopathology and the Origins of Agriculture*. New York: Academic Press.

Crosby, A. W. 1972. *The Columbian Exchange: Biological and Cultural Consequences of 1492*. Westport: Greenwood.

Davis, D. H., et al. 1975. Plague. In *Diseases Transmitted from Animals to Man*, ed. W. T. Hubbert, et al. Springfield, Ill.: C. C. Thomas.

Desowitz, R. S. 1980. Epidemiological-ecological interactions in savanna environments. In *Human Ecology in Savanna Environments*, ed. D. R. Harris. New York: Academic Press.

Dobyns, H. 1983. Their Numbers Become Thinned. Knoxville: University of Tennessee Press.

Dubos, R. 1965. *Man Adapting*. New Haven: Yale University Press.

Fiennes, R. 1978. *Zoonoses and the Origins and Ecology of Human Disease*. New York: Academic Press.

Fried, M. 1967. *The Evolution of Political Society*. New York: Random House.

Goodman, A., et al. 1975. The role of infectious and nutritional diseases in population growth. Paper presented to the annual meeting of the American Anthropological Association, San Francisco.

Hare, R. 1967. The antiquity of diseases caused by bacteria and viruses; a review of the problem from a bacteriologist's point of view. In *Diseases in Antiquity*, ed. D. Bothwell, and A. T. Sandison. Springfield, Ill.: C. C. Thomas.

Hubbert, W. T., et al. eds. 1975. *Diseases Transmitted from Animals to Man.* 6th ed. Springfield, Ill.: C.C. Thomas.
Hudson, E. H. 1965. Treponematosis and man's social evolution. *American Anthropologist* 67, 885-901.
Inhorn, M. C., and P. J. Brown. 1990. The anthropology of infectious disease. *Annual Review of Anthropology* 19, 89-117.
Jelliffe, D. B. et al. 1962. The children of the Hadza hunters. *Trop. Paed.* 60, 907-13.
Kent, S., et al. 1991. Hypoferremia: Disorder or defense. Paper presented to the annual meeting of the American Association of Physical Anthropologists.
Kochar, V. K., et al. 1976. Human factors in the regulation of parasitic infections: Cultural ecology of hookworm populations in rural West Bengal. In *Medical Anthropology*, ed. F. X. Grollig and H. B. Haley. Chicago: Aldine.
Livingstone, F. L. 1958. Anthropological implications of sickle cell gene distribution in West Africa. *American Anthropologist* 60, 533-62.
———. 1964. Human populations. In *Horizons of Anthropology*, ed. S. Tax, 60-70. Chicago: Aldine.
———. 1984. The Duffy blood groups, vivax malaria, and malarial selection in human populations: A review. *Human Biology* 56, 413-25.
Lovejoy, C. O., et al. 1977. Paleobiology at the Libben Site Ottawa Co. Ohio. *Science* 198, 291-3.
Marx, J. 1986. Viruses and cancers briefing. *Science* 231, 919-20.
McCullough, D. 1977. *The Path Between the Seas.* New York: Simon and Schuster.
McCullough, W. F., and J. S. Remington. 1975. Toxoplasmosis. In *Diseases Transmitted from Animals to Man*, ed. W. T. Hubbert, et al. Springfield, Ill.: C. C. Thomas.
McKeown, T. 1976. *The Modern Rise of Population.* New York: Academic Press.
McNeill, W. 1976. *Plagues and People.* Garden City, N.Y.: Anchor Press.
———. 1980. Migration patterns and infection in traditional societies. In *Changing Disease Patterns and Human Behavior*, ed. N. F. Stanley and R. A. Joske. London: Academic Press.
Miller, G. 1976. The epidemiology of Burkitt's lymphoma. In *Viral Infections of Humans*, ed. A. S. Evans. New York: Plenum Press.
Mims, C. 1980. The emergence of new infectious diseases. In *Changing Disease Patterns and Human Behavior*, ed. N. F. Stanley and R. A. Joske. London: Academic Press.
Mogabgab, W. I. 1976. Influenza. In *Communicable and Infectious Diseases*, ed. F. H. Top and P. F. Wehrle. St. Louis: C. V. Mosby.
Neel, J. V. 1977. Health and disease in unacculturated Amerindian populations. In CIBA, *Health and Disease in Tribal Societies.* New York: Elsevier.

Nelson, G. S., 1975. Schistosomiasis. In *Diseases Transmitted from Animals to Man*, ed. W. T. Hubbert. et al. Springfield, Ill.: C. C. Thomas.

Ortner, D., and W. Putschar. 1981. Identification of pathological conditions in human skeletal remains. Washington D.C. Smithsonian Cont. to Anthropology #28.

Ramenofsky, A. F. 1987. *Vectors of Death*. Albuquerque: University of New Mexico Press.

Rausch, R. L. 1975. Taeniidae. In *Diseases Transmitted from Animals to Man*, ed. W. T. Hubbert, et al. Springfield, Ill.: C. C. Thomas.

Rosen, M. N. 1975. Clostridial infections and intoxications. In *Diseases Transmitted from Animals to Man*, ed. W. T. Hubbert. Springfield, Ill.: C. C. Thomas.

Sanders, W. T., and D. Webster. 1978. Unilinealism, multilinealism and the evolution of complex societies. In *Social Archaeology*, ed. C. Redman, et al. New York: Academic Press.

Sattenspiel, L., and H. Harpending. 1983. Stable populations and skeletal age. *American Antiquity* 48, 489-98.

Scrimshaw, N. S., et al. 1968. Interaction of Nutrition and Infection. Geneva: WHO monograph 57.

Selby, L. A., 1975a. Blastomycosis. In *Diseases Transmitted from Animals to Man*, ed. W. T. Hubbert. Springfield, Ill.: C. C. Thomas.

———. 1975b. Coccidioidomycosis. In *Diseases Transmitted from Animals to Man*, ed. W. T. Hubbert. Springfield, Ill.: C. C. Thomas.

Service, E. 1975. *The Origin of the State and Civilization*. New York: W. W. Norton.

Sikes, R. K., 1975. Rabies. In *Diseases Transmitted from Animals to Man*, ed. W. T. Hubbert. Springfield, Ill.: C. C. Thomas.

Steinbock, R. T. 1976. *Paleopathological Diagnosis and Interpretation*. Springfield, Ill.: C. C. Thomas.

Stevenson, R. F. 1968. *Population and Political Systems in Tropical Africa*. New York: Columbia University Press.

Taylor, C. E. 1983. Synergy among mass infections, famines, and poverty. In *Hunger in History*, ed. R. I. Rothberg and T. K. Rabb. Cambridge: Cambridge University Press.

Weinberg, E. D. 1974. Iron and infectious disease. *Science* 184, 952-56.

PART THREE

The Industrial Era: New Societies, New Technologies, New Problems

4

The Revolution in the Family and the World We Have Made

David Levine

It is widely recognized that a critical component of the current set of problems labeled "environmental" is the rapid growth and urbanization of human populations that have occurred since the middle of the eighteenth century. When those rapidly growing numbers were coupled with the scale of production and consumption made possible by the industrial revolution, the kinds of global resource and environmental problems that are the focus of this volume emerged for the first time.

Most people familiar with the history of the impact of human beings on the earth date the emergence of global, massive, or devastating problems from the time of the demographic and industrial revolutions. And most people who study the issues involved focus on the aggregate level--that world population took all of human history until 1850 to reach 1 billion but added the next billion in just 80 years and the most recent, the fifth, in 12. Behind this admittedly staggering set of facts, however, lies a complex of social, economic, and cultural dimensions of those demographic changes. Of particular interest to us here are changes that the family underwent during the emergence of modern society in Europe, changes set in train by the demographic and industrial revolutions and culminated by the decline in European fertility that began in the last half of the nineteenth century.

The chapter that follows is devoted to consideration of two sets of ideas. The portion immediately following is a treatment of the political, economic, and social context giving rise to the changes in the family under discussion. That context is dominated by the shift from a predominantly rural European population engaged in farming, handicrafts, and non-mechanized community industries to a population more heavily urbanized and working for wages in mechanized factories fueled by inanimate sources of energy. The latter portion of the chapter considers the sorts of political changes that accompanied the shift in the family. Chief among these changes are the development of the modern nation-state, the emergence of the notions that working-class families had too many children and that it was the province of the state to induce working-class parents to limit their families, and the development of compulsory state-run public education. In sum, the organization of national social systems during the later nineteenth century provided the historical context in which the revolution in the family was keynoted by the decline in fertility. This important change was both an innovation and an adjustment, not only responding to large-scale changes in social organization but also representing one of the original ways in which individual men and women acted to make their own history.

The century before 1770 was one of rough demographic stability; it was followed by a period of rapid growth that would, by the end of the eighteenth century, create a whole new species of urban life--indeed, a kind of life entirely different from what had preceded it. At this time, the great mass of the population was in the last throes of proletarianization; they were becoming members of an urban working class. A way of life was being lost as they were becoming wage-earners and concomitantly losing their purchase on the land. The value of women's and children's labor was radically depreciated; they were marginalized by the new political economy of urban and industrial capitalism. In moving towards the patriarchal breadwinner economy, the social standing of those who were neither patriarchs nor breadwinners declined. Individual peasant families struggled desperately against proletarianization and against the larger historical forces that gave these struggles an air of quiet desperation. Demographic behavior became only one of a series of coping maneuvers within the calculus of conscious choice. Within the changing tempo of daily life during this age of revolution, family formation strategies were both conceived anew and put into effect.

In most places, the period after the middle of the eighteenth century was one of rising population in response to falling levels of mortality and roughly stable levels of fertility (Flinn 1974, p. 3). But declining mortality rates in turn released uncontrollable forces when unexpected levels of survival--of youths and married adults--combined with earlier marriage and skyrocketing illegitimacy rates. Married couples remained intact as units of demographic reproduction for longer periods of time while a much higher proportion of children was likely to reach adulthood and themselves marry. Urbanization, with its filthy environment breeding microorganisms so lethal to babies, partially counterbalanced the global improvements in life expectation (see Cohen, this volume). Overall, however, mortality rates dropped and, as we know from simple Malthusian arithmetic, even small changes when aggregated and allowed to multiply over several generations had profound implications.

Other effects resulted from this shift in the mortality schedule; not the least was the changing configuration of the age-pyramid, which rapidly broadened at its base. Better chances of child survival combined with the diminishing chance of marital break-up to swell the lower age groups at the end of the eighteenth century. In itself this factor played no small role in contributing to the maintenance of high aggregated fertility rates. Generations followed one another more quickly. Villages of the early nineteenth century reached their peak population totals as the social system inhaled these surplus bodies and attached them to precarious niches in the local economy. The countryside filled up quickly. Dwarf-holdings multiplied, intensive cultivation and new crops along with a more vigorous division of labor in the service sector combined to make it possible for the land to fill up to the point of super-saturation. It could not continue; it did not.

The second half of the century witnessed unprecedented levels of migration, both external--millions and millions and millions of people left Europe, most of them after mid-century--and internal--continental urbanization proceeded furiously to catch up with British levels where forty per cent of the population lived in six large conurbations by 1881 (Hobsbawm 1968, p. 131).

The forces holding villagers to their *pays* or *heimat* were in tatters. Many tried desperately to stay, many left in desperation. By seeing strategies of persistence and strategies of migration together, we can begin to understand how the demographic revolution catalyzed the destructive fury unleashed by the industrial revolution, and *vice-versa*. Together they wreaked a veritable holocaust on an overstretched peasantry, desperately clinging to cottage industry as a supplement to their subdivided holdings, by undermining the continued viability of the family farm and the family production unit. The family production unit's reliance on its own labor power merely served to expose it, nakedly, when the terms of trade swung violently against it in the mid-nineteenth century. The rise of factory-based cotton spinning led to massive unemployment of women and children. This story was repeated throughout the Industrial Revolution and across the face of rural Europe in the nineteenth century.

During this time, technological change continuously modified the frontiers of what was humanly possible in the age-old quest to master the material world. Moreover, change built up its own momentum. The quantum leap afforded by the substitution of inanimate, inorganic energy for natural energy was equivalent to creating an industrial reserve army of slaves. This metaphor, suggested by E. Levasseur at the end of the nineteenth century, relates to the French economy between the 1840s and the 1880s when the augmentation to the labor force was equivalent of 98 million new hands--or, as he put it, "deux esclaves et demi par habitant de France" (quoted in Wrigley 1988, p. 76). France brought up the rear among the leading industrial economies--behind England, Germany and the American colossus. Just imagine the combined size of this industrial reserve army of slaves!

The history of energy consumption and the labor process provides one grammar of the industrial revolution; the history of human life provides another. Thomas Hobbes, writing in the mid-seventeenth century, noted, in one of the most famous and misquoted historical observations of the early modern age, "And the life of man [sic], solitary, poore, nasty, brutish, and short" (Hobbes 1968, p. 186). Hobbes was no demographer, but research in that field has amply substantiated his point. Quite simply, people born in 1750 lived shorter, less healthy lives than we do now. Put schematically: two in ten children died in infancy; another two of the original ten died before reaching puberty; still more either died before marrying or never married. Those who did survive and marry spent most of their adult lives either reproducing themselves or raising the next generation. Life-cycle stages were fewer in number. People born in 1750 would expect to die about twelve years before the birth of their first grandchild; we usually live twenty-five years after the birth of our last grandchild (Anderson 1985, p. 72ff).

The industrial revolution, in time, changed the dreadful conditions of the early modern world. With the aid of a revolutionary technological break from the age-old dependence upon animate sources of energy, capitalism was successful in providing a growing population with both basic necessities and more consumer goods of seemingly unimaginable variety. Everyone living today in an advanced capitalist society benefits from this form of industrial slavery although, to be sure, such benefits have been divided unequally, and the trickle-down effect has been hardly immediate. But there can be no doubt that more people living in advanced industrial societies now have more--and live longer and healthier lives--than did their predecessors.

The transition between the misery of the early modern period and the relative plenty of modern times provided the historical context for declining fertility. Since the middle of the nineteenth century, fertility in Europe has fallen, and since 1870 the "stopping pattern" has become the primary method of fertility control. Before this time the "starting" and "spacing" patterns had previously been employed, albeit with varying degrees of success, as regulatory strategies (Lesthaeghe 1980, pp. 527-548). In the third quarter of the nineteenth century, "stopping" was exceptional; four generations later this pattern had become almost universal. The transformation of marital fertility has been both rapid and complete. The difference between marital fertility practices in 1870 and those which prevail today represents a dramatic change in lived experience.

That change followed the revolutionary transformation in European economic, political, and social conditions already described. The nineteenth century witnessed the final stage of this process of transformation. As Robert Muchembled writes, "One model of society replaced another." He goes on to describe the organization of the world we have lost in the following terms.

> A 'polysegmentary' society. . . a social system composed of many subgroups: clans, lineages, families, and kinship relationships, age groups, corporations, guilds, fraternities and confraternities, parishes and neighborhoods, village or urban communities, and so on. The reality of existence and the organization of power were situated at the level of these subgroups and of their organization. [For example], France was made up of thousands, even hundreds of thousands of politico-domestic units that overlapped and that reached their greatest cohesion in the framework of the village and of the urban quartier, rather than in that of an entire city or town, a region or a province (Muchembled 1985, p. 313).

This model of society, while still existing vestigially in the interstices of modern society, came under acute attack in the era of the demographic transition. Spatially, Europe was redrawn; the patchwork of horizontal bonds--Muchembled's "politico-domestic units"--was reconstituted and replaced by a small number of vertically-integrated nation-states. The immense growth and influence of the nation-state was achieved at the cost of lesser sovereignties. In the era of declining fertility, national differences became more important than regional or local ones.

The nineteenth century was an age of revolution, of the bourgeoisie, and of empire. The mid-Victorian boom drove the entire world into the capitalist political economy. And, at the same time, it precipitated yet another ratchet-turn in the formation of national state structures superseding the parcellized sovereignty of the medieval world and its "polysegmentary" social systems. The nineteenth-century contribution to the process of state formation hinged on merging the traditional goals of social discipline with the peculiar demands of mass society. The key institution in cementing this merger was the classroom, where pedagogy impregnated the young with a civic catechism in a national language (see Hobsbawm 1988, pp. 149-150; Corrigan and Sayer 1985). This new pedagogical project formed human capital into citizens of the state. At the heart of every nineteenth-century curriculum was the study of national history, which cultivated civic consciousness and promoted patriotism in a world where human destinies were seen to be manifest in the actions of the state rather than immanent, in the word of the gospel. Modern, state-directed discipline inculcated obedience to bureaucratized routine while legions of functionaries implanted this habit into everyday life. In place of the great chain of being, modern society was brought to observe the rhythms and conventions of the nation-state based upon its newly-invented traditions of citizenship and nationalism.

The key intersection between state formations and family formations took place at the points of reproduction: demographic, cultural, social and political. Michel Foucault writes that central to the economic and political issues raised by this intersection was sex. "Sex was not something one simply judged; it was a thing one administered." It was not intractable, but rather was endowed with "instrumentality". Its specific domain was "the family cell" whose "normalization" was based on "a deployment of sexuality": "a hysterization of women's bodies"; "a pedagogization of children's sex"; "a socialization of procreative behaviour"; and "a psychiatrization of perverse pleasure" (Foucault 1980, pp. 25, 24, 103, 108, 104-5). Sex, for Foucault, is not simply eroticism, but rather the biological reproduction of social structure--what he calls "bio-power"--in all its dazzling multiplicity of possibilities. It was to be administered--or, as he would say, "normalized"--in order to privilege the centripetal demands of the social system over the centrifugal demands of the individual.

Foucault locates the emergence of "bio-power" with the political arithmeticians of the mercantilist age. They sought to administer states, not just rule them. From an original concern with quantity, nineteenth-century administrators became increasingly interested in population quality and proactive in their attempts to foster it.

It was in response to proletarian proliferation that moralizing Malthusians focused on disorderly family government as the correlative of indigence and poverty. Malthus's own writing had an electric effect: he not only defined the moral problem; he also formed the discourse on the subject of poverty and social policy. While there was no unanimity in support of his views and, indeed, a considerable body of elite paternalists were revolted by them, nonetheless Malthus gave the issue of proletarian proliferation "a centrality it had not had before, made it dramatically, urgently, insistently problematic" (Himmelfarb 1983, p. 126). Malthus did this by connecting the older ideas of "police"--the maintenance of hierarchy, morality, and social order--with newer concerns regarding the statistical organization of public life.

The French Revolution transformed the politics of police amidst an atmosphere of upper-class fear and loathing. Not coincidentally, Malthus's tract was written for its times: aimed squarely at the utopianism of Condorcet, anglicized by William Godwin, Parson Malthus did not want to force men and women to be free; instead he wanted to force them to be constrained by their own inner compulsion and common sense. Malthusian common sense was not the *Common Sense* of Thomas Paine; rather it was based on an understanding of the real world. This is why Malthus could argue that profligate, undiscriminating charity--that of the Poor Laws--ran contrary to the rules of "nature's mighty feast." Malthusian social policy has ever since been based on the understanding that the disorderly family was the problem while the ordered family was the solution to the population question (Andrew 1989, p. 168-169). In Lille, Mulhouse and Rouen, as in the English coal mining region of County Durham, the proliferation of proletarians disturbed social policy-makers.

When the echoes of the *Marseillaise* subsided, nineteenth-century Malthusians regarded the menacing features of urban, industrial society to reside "in a realm of human behavior removed from the public theaters in which working-class challenges to the status quo usually took place" (Lynch 1988, p. 232). So, the personal was politicized. The working-class family's profligacy--its population power--was thus the outer, public face of its inner, private lack of control. Proletarians' high fertility was incomprehensible to the bourgeoisie who considered their additional children to be mouths to feed while the working class considered them to be hands to work and to earn and to insure the family against the ill-luck of any particular member. This distinction between the political economy of individualism and the moral economy of mutuality found deep resonances in class-specific family formation practices (see Smith and Valenze 1988, pp. 277-298). Malthus's legacy, in France as well as in England, was to have castigated the moral economy of the proletarian family as evidence of its deficient moral education. Policing the working-class family was, then, the moral equivalent of the class war.

Domesticity provided the key to controlling proletarian fertility. In contrast to the rough immorality of the street, female mentalities were re-educated to make the cozy hearth a proper home. Work was reclassified as a masculine endeavor; masculinity was tested in the labor process and judged by the harmony of domestic discipline and its independence. Propriety was to be the moral property of women, who were reformulated as model wives and mothers.

> The vice of the bad home, and the virtue of the good home, could be just as easily translated to descriptions of the good and bad wife: one was chaotic, promiscuous, unsettled, sensual, dirty, and unhealthy; the other was orderly, modest, stable, rational, clean, and well. And what was true of the wife in her home determined the family also (Colls 1987, p. 137, 144).

In the process of legitimating its own social vision, the bourgeoisie located social dysfunction in the proletariat's estrangement from the respectable family mores of propriety and independence. Lynch writes as follows.

> This powerful sense of moral community at the family level stemmed from what was seen as the family's capacity to mediate human needs for affection and individual self-realization with the necessity of providing predictable material resources to manage those aspirations concretely within a society whose recent memory was marked by the specter of social and political disorder (Lynch 1988, p. 226).

To achieve the moralization of the proletarian family, it was necessary to legitimate working-class marriages; this was done by integrating consensual unions and their out-of-wedlock children into respectable society--though often by removing children from their parental home--and encouraging the virtues of thrift and decency. Legitimating these children may have been the solution; their abandonment was the problem. This problem was thought to reflect a feckless attitude to sexual unions; if, by some chance, a stable union was formed in the bowels of the factory city, then there seemed to be every likelihood that fecklessness would characterize parent-child relations. Such children, it was claimed, would be enslaved to their parents who would sell them into wage-slavery. In the name of protecting proletarian children from their parents, reformers were willing to abrogate parental (and especially paternal) rights.

Institutional surveillance and school attendance were the prescribed regimen for the moral education of the younger generation. Families would be policed, and the free time of youngsters could be supervised in a setting that was consciously constructed in opposition to the popular culture of the street-corner, the tavern, the factory and the home. Schooling (and then domestic service for older girls) would inculcate this official sexual ideology in order to complete a favorable circuit of class relations.

For the proletariat, then, the deployment of sexuality was part of a *mission civilatrice* in which bourgeois missionaries sought to convert them ("normalize" them in Foucault's terms) by means of the following.

> . . . an entire administrative and technical machinery. . . which made it possible to keep that body of sexuality, finally conceded to them, under surveillance (schooling, the politics of housing, public hygiene, institutions of relief and insurance, the general medicalization of the population. . . (Foucault 1980, p. 126).

The emergence of sexual politics occurred when, in the pithy words of Jacques Donzelot, "*morality was systematically linked to the economic factor*, involving a continuous surveillance of the family, a full penetration into the details of family life" (Donzelot 1979, p. 69, emphasis added). In contrast to the "government of families" that occurred in the early modern period, in the nineteenth century domestic politics came to pivot on "government through the family." Donzelot (1979, p. 92) argues further:

> [T]he modern family is not so much an institution as a *mechanism*. . . . A wonderful mechanism, since it enables the social body to deal with marginality through a near-total dispossession of private rights, and to encourage positive integration, the renunciation of political right, through the pursuit of well-being.

The discipline of the body, the mind and the soul were part of the "infraculture" of western history. Its diffusion, and more particularly its inculcation, were at the heart of that "long and arduous process of education"--a cultural capital whose accumulation was hardly primitive-- which so impressed Max Weber (Weber 1958, p. 62). Foucault joins the great German sociologist in believing that the roots of this process stretched back beyond the industrial revolution to the Reformation when the classical Christian pastoral message was once again democratized after a millenium of clerical monoploy. Even before the Reformation, the western Christian confessional technology of private self-examination played a central role in the order of civil and religious powers. In the course of the early modern period this technology of the self was severed from religion; self-examination became popularized as the real way of life, not just a formalized set of ritual observances in the charge of specialists.

> It is to be noted that to him that is a governor of a public weal belongeth a double governance, that is to say, an interior or inward governance, and an exterior or outward governance. The first is of his affects and passions, which do inhabit within his soul, and be subjects to reason. The second is of his children, his servants, and other subjects to his authority (Hodgkin 1990, p. 21).

Foucault diverges from Weber not in his interest in worldly asceticism but rather in his own way of connecting regimens of repression to secular systems of domination and power. The nineteenth-century state was able to bureaucratize the policing of families by incorporating it into the daily life of its citizens--from cradle to grave--and by cloaking it in the postivistic mantle of medical science. The regulatory thrust of its command structures was built upon deep foundations; in this regard, the nineteenth-century state formation was modern in that it was able to avail itself of new institutions, new techniques and new technologies. In place of the sacrament, secular authority placed its faith in the family crystal, the "Malthusian couple" (Foucault 1980, pp. 105-111). All the forces of the combination of power and knowledge--or "power/knowledge," as Foucault styles it, that stands for the sum of all authority and coercive or repressive capacities of the state--come to bear on the family.

The family axis became both more narrowly construed and more attentively policed. It became the site for sentiment. The most rigorous techniques of repressive discipline "were formed, and, more particularly, applied first, with the greatest intensity, in the economically privileged and politically dominant classes." Foucault writes that "what was formed was a political ordering of life, not through an enslavement of others, but through an affirmation of self."

> [I]t has to be seen as the self-affirmation of one class rather than the enslavement of another: a defense, a protection, a strengthening, and an exaltation that were eventually extended to others--at the cost of

different transformations--as a means of social control and political subjugation (Foucault 1980, p. 120).

Resistance to this new disciplinary project was based on something more than an irrational adherence to tradition. Only by adopting the political stance of the socially dominant--that is, older male members of the bourgeoisie--can we make the ideological leap and join them in castigating resistance to change as unthinking and hence irrational. Looking at social change from the top down, however, is no way to get inside the survival strategies employed by the mass of the population. Family formation strategies were not reflex actions; they represented something deeper that adapted to changing pressures by assimilating what they needed and rejecting the rest. In an age of revolution, however, the pressures to adapt became more intense and the resistance to change more complex. In place of independent paternal authority, which characterized the early modern family--and provided the organizing metaphor for its political theory--the later modern family was defined by the modern state apparatus permeating its domestic space.

Coincident with the fertility decline, then, a new disciplinary complex congealed; orchestrated by the state, the helping professions coordinated the application of "power/knowledge" in the daily lives of the citizenry. Teachers, social workers, psychologists, and the whole battery of social welfare agencies were created from a patchwork of voluntary institutions that had previously been called upon to aid the sick and infirm. Under the mantle of professionalization, the job of interpretation devolved into the province of a new technology of the self staffed by "the directors of conscience, moralists and pedagogues". Foucault writes that Freud's articulation "of the Oedipus complex (as an explanation for neurosis) was contemporaneous with the juridical organization of loss of parental authority" (Foucault 1980, pp. 128, 130). This reconstituted family became not just the locus of "bio-power" but also, and inevitably, the apparatus through which the practical regimen of reproduction was carried out. The reproduction of "bio-power" was thus both the means and the end of the revolution in the family. Foucault's insistence on the nature of power relations in the politics of family formation provides an important dimension to the processes of social change within which the decline of fertility occurred.

The hallmark of this new knowledge-politics was the state's enhanced power to police and invigilate, discipline and punish, and reward. The growth of a state apparatus concerned with human capital speaks much very to this point. Indeed, it provides the glue which joins together the public and the private, the social and the individual. We can locate its beginnings in the statistical study of social relations in which

> . . . the flow of information about the lower classes [w]as the countermovement of the flow of domination from the top of the social structure to the bottom. . . . [S]ocial domination increasingly required accurate information flowing upward to social and political leaders (Lynch 1988, p. 196).

The fundamental sites of this new disciplinary program were the public school and the private family. Faced with recalcitrance and outright resistance, social disciplinarians sought recourse to the courts and argued that it was both a social and an individual good to break up immoral family units. This was the field of moral force within which the massive expansion of compulsory schooling orbited.

In the school--with its hidden curriculum favoring obedience and rote learning as well as its gendered differentiation of the fit and proper concerns for the 'education' of boys as opposed to girls--control over the socialization of the young was wrested from the family and the community in order to be made into an exercise in administration, categorization and subordination. Compliance with the rule of law in modern societies is a reflection of the way in which "a gentle empire of moral habits" (Lynch 1988, p. 79) exercises suzerainty over the everyday life of citizens. This project of moralization was the political terrain on which compulsory education was erected. Edmond Holmes, a Victorian school inspector, recollected the following.

> For me they were so many examinees and as they all belonged to the 'lower orders' and as (according to the belief in which I had been allowed to grow up) the lower orders were congenitally inferior to the 'upper classes' I took little or no interest in my examinees either as individuals or as human beings, and have never tried to explore their hidden depths. Indeed, the idea of their having hidden depths was foreign to my way of thinking, and had it ever presented itself to my mind I should probably have dismissed it with a disdainful smile (Holmes, 1920, p. 64).

With the benefit of hindsight, Holmes was to regret the moralizing pedagogy provided to the "examinees" in the state's compulsory schools. It was, he came to understand, "in the highest degree anti-educational": the average fourteen year-old would have been subjected to 2000 "scripture lessons" and 3000 "reading lessons" in the course of her or his "acutely de-vitalizing" schooling (Holmes 1914). School-leavers--the product of a moralized family and a moralized pedagogy--were given a certificate and sent away to enter the labor market and national armies.

The progress of mass education formed and shaped young minds to become citizens of a new moral world. That moral world, like all others, was predicated on the institutionalization of power relationships and, of course, the marginalization of those without power. The powerless learned to suffer and be still. They were not, however, economically deprived, for it is the genius of modern capitalism that producers must also be consumers, and in precisely this way the fetters of production have been replaced by those of consumption. Most members of modern society, like Gulliver awakening in Lilliput, have been tied to the consumer economy by a thousand tiny chains. The political economy of this modern intellectual adventure has trained mass populations to be obedient consumers--of mass education, mass politics, and mass production. In this regard, the formidable role of the war machines of rival imperialist powers to regiment their populations wildly exceeded anything envisaged by earlier political arithmeticians. The creation of command economies, orchestrated by military-industrial complexes, meant that "the transcontinental integration of human effort attained by the 1870s constituted a landmark of world history" leading to "the twin processes that constitute a distinctive hallmark of the twentieth century: the industrialization of war and the politicization of economics" (McNeill 1982, pp. 261, 294).

The reproduction of human capital draws this discussion to the final dimension of contingent, historical time so as to consider intimate interactions structured by gender, age, the life-cycle, and family life itself. The reproduction of human capital connects human agency with the large-scale processes of manufacturing consent. Fertility control was not possible without the active involvement of both husbands and wives; the widespread prevalence of abortion stands in testimony to the studied indifference of men. But the sentimentalization of the family made it more likely that methods of control would come within the parameters of conjugal agreement. Because most of the decline in marital fertility was the product of two ancient practices--abstinence and *coitus interruptus*--it is clearly important to pay attention to the changing temper of communication between husbands and wives. Cultural forces, especially the regendering of the marital union, both narrowed the scope of family life and intensified its internal dynamics.

> In order to understand the cautious prudery, the well-controlled life-style, the emphasis on the home and its order in contrast to the outside

world and its chaos, it is important to see these phenomena for what they really were: a defense mechanism against the social unrest that was plaguing society (Frykman and Lofgren 1987, p. 259).

Demographic statistics are testimony of profound social change. Accompanying the industrial revolution, urbanization, the doubling of life expectation, the institutionalization of mass society, the politicization of war and the militarization of the world economy, the transformation of reproduction patterns was part of a massive shift in the nature of social relations. The effects of this revolution are visible in the statistics of fertility and mortality as well as in the reconstruction of gender and the newly-reconstituted life-cycle, with its sequential age-graded roles and activities, providing the framework within which we need to locate the making of the modern family. And, furthermore, the causal arrows also flowed in the other direction: changing forms of behavior in turn modified social systems. It is in this sense that the texture of intimate relationships was not just important in its own right but also of crucial significance in reconstructing attitudes towards fertility. This is not simply a matter of acknowledging yet another independent variable so much as shifting our focus towards the culture of fertility. The demographic implications of these changes were recognized by Thorstein Veblen who argued that "the low birthrate of the classes upon whom the requirements of reputable expenditure fall. . . is probably the most effectual of the Malthusian prudential checks" (Veblen 1967, 113). Moreover, as James Wickham has argued about proletarian family life in Weimar Frankfurt, "The ideology of the respectable family, nucleated and domesticated, was the private property of neither left nor right" (Wickham 1983). Both parents-- the respectable husband and the Malthusian wife--decided together to control their fertility in order to enhance their domestic respectability and to improve the life-chances of their children. In place of the economically useful child, the modern family has substituted the emotionally priceless one or two (Zelizer 1985).

The proletarian family was revolutionized in the course of the transition from early modern to modern times, when the proletariat became the overwhelming majority in the European population. The working-class household became the site of social and biological reproduction, not production, and in the process it came to be judged by the quantity and the quality of its product--i.e. human capital. We can see the pre-history of this transformation in the debates on police and charity initiated by the early modern political arithmeticians and political economists, but it was only two centuries later that the institutional instruments were put in place to realize this Malthusian positivity, the modern family. Before that happy day came to pass, many of the social functions of the working-class family--education, health and welfare--were superseded by the aggressively intrusive actions of the modern state while its productive functions were redefined by industrial capitalism. The re-structuring of social relations of production was intimately connected with the implosion of the working-class family--as a result of the exclusionary tactics of trade unionization and a renewed patriarchy which privileged adult males. The restriction of working-class fertility was another tactic by which the labor supply was controlled when the cost of its reproduction began to soar as children were removed from productive, waged work and kept in school by the rules and regulations of the state. It was only in these radically transformed circumstances that it became realistic for the working-class family to consider modernizing itself.

In the face of the prescriptions urged upon them by respectable moralizers, the working-class family organized itself. We can see the impact of this transformation most clearly in the changing life-cycles and gendered roles of its members. In contrast to the corporate life-cycle of its predecessors, the urban proletariat was decidedly modern in that its children stayed at home until marriage. Marriage was itself predicated on the ideal of the male breadwinner and the domestication of the wife-mother. The commodification of labor power, and the concomitant devaluation of skill, compressed the life-cycle of the proletariat. Instead of a two-phased transition from childhood to youth and then from youth to adulthood--the first marked by leaving home and the second by marriage--which characterized the corporate life-cycle (as well as its rural equivalent of service), the proletarian life-cycle was marked by a single transition in which leaving home and marriage were squeezed together.

The huddling of working-class families was necessitated by the disintegration of older modes of production and compounded by the massive increase in the supply of labor occasioned by the demographic revolution. On the one hand, adult male workers lost control of the labor process while on the other hand they lost control over their patriarchal family capital. In these circumstances, it was possible for only a tiny minority of labor aristocrats to approximate the gendered breadwinner ideal urged upon them by the dictates of respectability and bourgeois moralizers before the end of the nineteenth century.

The transformation of demographic behavior was strongly influenced by the way in which those who had been marginalized by the experience of pre-industrial modernization made their own history. It was their pro-activity--expressed in demographic terms by their more precocious ages at first marriage and expressed in social terms by their rising expectations for inclusion in both civil and consumer society--that transformed the task of achieving prosperity. This task was further complicated because the relative openness of the early modern marketplace was challenged by a mean-spirited class-consciousness sparked by the Malthusian project of blaming the poor for their poverty. Yet, the openness of the marketplace survived; so, too, did the responsiveness of the political system. Social dislocation was generated by the proletarianization of those who had been marginalized in the rise of the middle class. That marginalization resulted from the intensification of the contradictions of early modernization that could promise inclusion but was predicated on the exploitation of one half of society for the liberation of the other half. Modern consumer culture is now within the reach of everyone and the grasp of most. Mass society is built on the foundations laid down in the pre-industrial period when early modernization was achieved at the cost of restricted access. The working-class family modernized not so much by becoming bourgeois as by converging around a certain normative standard acknowledged by all though interpreted in quite distinctive ways according to class, ethnicity, gender, and age.

Today, the loss of that prescriptive unanimity is a matter of fact. Those who mourn that loss cannot forget the family; they seem to believe that it was a natural form of social organization and not one constructed in the transition from early modern to modern times. The cost exacted by modern memory is that those who cannot forget the family often mistake its appearance and in so doing betray a fundamental misunderstanding of the contingengy of the world we have made.

References

Accampo, E. 1989. *Industrialization, Family Life and Class Relations: Saint Chamond, 1815-1914* Berkeley and Los Angeles: University of California Press.

Anderson, M. 1985. The emergence of the modern life cycle in Britain. *Social History* 10, 69-87.

Andrew, D. 1989. *Philanthropy and Policy: London Charity in the Eighteenth Century.* Princeton: Princeton University Press.

Colls, R. 1987. *The Pitmen of the Northern Coalfield.* Manchester: Manchester University Press.

Corrigan, P., and D. Sayer. 1985. *The Great Arch.* Oxford: Basil Blackwell.

Donzelot, J. 1979. *The Policing of Families.* New York: Vintage Books.

Flinn, M. W. 1974. The stabilization of mortality in pre-industrial Western Europe. *Journal of European Economic History* 3, 285-318.
Foucault, M. 1980. *The History of Sexuality*, V. 1, An Introduction. New York: Vintage Books.
Frykman, J., and O. Lofgren. 1987. *Culture Builders: A Historical Anthropology of Middle-Class Life*. New Brunswick: Rutgers University Press.
Harrison, J. F. C. 1981. *Early Victorian England 1832-1851*. London: Fontana Books.
Himmelfarb, G. 1983. *The Idea of Poverty*. New York: Vintage Books.
Hobbes, T. 1968. *Leviathan*. C. B. Macpherson. ed. Harmondsworth: Penguin Books.
Hobsbawn, E. J. 1968. *Industry and Empire*. London: Allen Lane.
Hodgkins, K. 1990. Thomas Whythorne and the problems of mastery. *History Workshop* 29, 20-41.
Holmes, E. 1914. *What Is and What Might Be*. London: Constable.
———. 1920. *In Quest of an Ideal*. London: R. Corbin-Sanderson.
Ladurie, E. L. 1974 [1966]. *The Peasants of Languedoc*. Urbana: University of Illinois Press.
———. 1981. *The Mind and the Method of the Historian*. Chicago: University of Chicago Press.
Lesthaeghe, R. 1980. On the social control of human reproduction. *Population and Development Review* 6, 527-48.
Lynch, K. A. 1988. *Family, Class and Ideology in Early Industrial France: Social Policy and Working-Class Family 1825-1848*. Madison: University of Wisconsin Press.
MacNeill, W. H. 1982. *The Pursuit of Power*. Chicago: University of Chicago Press.
Muchembled, R. 1985. *Popular Culture and Elite Culture in France 1400-1750*. Baton Rouge: Louisiana State University Press.
Samuel, R. 1977. The workshop of the world: Steam power and hand technology in mid-Victorian Britain. *History Workshop* 3, 6-72.
Smith, R. L., and D. M. Valenze. 1988. Mutuality and marginality: Liberal moral theory and working-class women in nineteenth century England. *Signs* 13, 277-98.
Veblen, T. 1967. *The Theory of the Liesure Class*. New York: Viking Press.
Weber, M. 1976. *The Protestant Ethic and the Spirit of Capitalism*. New York: Charles Scribner's Sons.
Wickham, J. 1983. Working-class movement and working-class life: Frankfurt-am-Main during the Weimar Republic. *Social History* 8, 333.
Wrigley, E. A. 1988. *Continuity, Chance and Change: The Character of the Industrial Revolution in England*. Cambridge: Cambridge University Press.
Zelizer, V. 1985. *Pricing the Priceless Child*. New York: Basic Books.

5

Pollution and the Emergence of Industrial America

Martin V. Melosi

Industrialization of the United States during the nineteenth and early twentieth centuries intensified the impact of energy use on the environment. The transition from a wood-based economy to one dependent on fossil fuels resulted in acute air, water and land pollution. The newly emerging economic order also led to the concentration of factories and workers in urban areas, which contributed to existing environmental problems. While local impacts predominated in these years, they became chronic in every part of the country touched by industrialization.

The first major efforts to seek remedies for industrially induced pollutants also can be traced to this time. Indeed, the debate over economic progress versus environmental protection--so much a part of modern political rhetoric--is but the most recent phase of a controversy more than 100 years old.

History of U.S. Industrial Energy Use

Starting in the Middle Atlantic and North Central states, several factors contributed to the rise of large-scale industrialization. New machinery replaced hand tools and muscle power in the fabrication of goods, and the harnessing of steam power allowed production to be centered in urban-based factories. While mechanization undercut the need for workers with strong backs, it required a large labor force with the agility to operate equipment. An organizational revolution led to better coordination between management and production and encouraged the formation of large, integrated companies, capable of exploiting regional and national markets. State and federal governments actively promoted industrial development. And westward expansion provided crops for export and precious metals for increasing the money supply (Klein and Kantor 1976, pp. 4-6; Russel 1964, pp. 186-99).

While agriculture accounted for the largest share of production income before the Civil War, manufacturing rose rapidly during the following decades. In 1859, the United States had 140,000 industrial establishments--many hand or neighborhood industries. Just forty years later, there were 207,000--excluding hand and neighborhood industries. By 1900, the United States was the world's leading manufacturing nation (Degler 1967, pp. 31, 42-43; Russel 1964, p. 338).

The conversion to fossil fuels accelerated the industrialization process, helping to transform the United States into a modern nation. The exploitation of natural resources also led significantly to the deterioration of the environment (Melosi 1985, pp. 17-18).

Fuelwood has the longest history of any energy source in the country. Important in heating and cooking, wood also helped to spur the development of locomotives and steamboats, the charcoal-fired iron industry, and the stationary steam engine (Hindle 1975, pp. 3-5; Taylor 1951, pp. 56-73; Cole 1970, pp. 355-56; Burlingame 1946, pp. 193-214; Petulla 1977, pp. 102-3, 122-24).

Waterpower also was vital for some industrial uses. "Although the basic elements remained much the same, . . ." Louis Hunter stated, "the scale, complexity, and refinement of details in design and operation found in such major hydropower installations as those of the New England textile centers bore slight resemblance to the water mills in which they had their origins." The shift of manufacturing to cities undercut the use of waterpower, but its earlier vitality and that of wood suggests the value of renewable sources in the nineteenth century (Hunter 1979, pp. 159-69, 181-84, 188; see also von Tunzelmann, this volume).

Coal soon became the preferred fuel of the industrial era. Dominance of fossil fuels was due less to depletion of wood or waterpower than the ability of coal, and later petroleum, to better adapt to mechanization than the bulky, immobile and less versatile renewable resources. Dependence took years, but after barriers were overcome, coal grew steadily in importance.

Alfred Chandler (1972, pp. 72-101) contended that anthracite (hard coal) from eastern Pennsylvania provided the basic fuel for power and heat in the urban factories at mid-century. The availability of anthracite, advances in steam power, and new supplies of iron led to growth in several industries. "Why," he queried, "did factories, which had become significant in British manufacturing by the end of the eighteenth century, not become a major form of production, except in the textile industry, in the United States until the 1840s?" He concluded that the availability of more and better iron through the use of coal, in combination with greater use of steam power, changed manufacturing in the 1830s. Early textile mills, powered by water and equipped with machines made of wood and leather belting, were replaced by factories powered by steam, equipped with metal machinery, and located in cities.

But changes in iron production provide only a partial answer to dependence on coal. Americans first made major use of anthracite as a lighting source during a fuel crisis in the Philadelphia of the War of 1812. Although residents in the anthracite region of Pennsylvania had used local hard coal before the war, Philadelphians relied on bituminous (soft coal) from Virginia and Great Britain. The British blockade reduced those supplies, forcing consumers to obtain coal from the anthracite region (Powell 1980, pp. 5-10).

Bituminous coal as an industrial fuel soon surpassed anthracite. Soft coal became popular because of its availability, versatility, heat content, and compactness. The use of the Kelly-Bessemer process in the 1860s, and later the open hearth process, made bituminous a key ingredient in steel and iron production. By World War I, its sales outpaced anthracite by 450 percent (Kirkland 1969, pp. 302-4; Petulla 1977, pp. 178-80).

Coal achieved dominance because it was adaptable to more than industrial needs. By the 1830s, anthracite produced steam for factory machinery, steamboats and locomotives. Railroads, especially, made coal preeminent because it became the primary revenue source (Tarr and Koons 1982, p. 72). Coal invaded the domestic heating and lighting markets as early as the 1820s. At its peak in 1920, more than 658 million tons were mined in the United States. Consumption increased seventy-seven times between 1850 and 1918 (Binder 1958, pp. 83-92; Schurr and Netschert 1960, pp. 66-74).

The petroleum industry, heir to the energy crown in the mid-twentieth century, was born in the age of coal. Like its fossil fuel counterpart, petroleum began as a specialized form of energy, first emerging as a source of artificial light in the late-nineteenth century, then as a superior lubricant, and finally as the leading transportation fuel. The oil industry had its roots in the heart of Pennsylvania coal country. But its most dramatic impact grew out of strikes in the Southwest and West in the twentieth century. The seemingly endless supplies in fact undercut the economic dominance of the Northeast and Midwest as mechanization crossed the Mississippi.

Until World War I, increasing demand for oil did not threaten coal as the nation's leading energy source. Yet the importance of oil grew steadily. Local successes eventually enlarged the market for oil and led to improved transportation, conversion of coal-burning equipment, and better marketing. While electricity displaced kerosene, the automobile turned petroleum refining toward gasoline. With total energy consumption more than doubling between 1900 and 1920, oil became significant in absolute terms, if not as a percentage of total consumption.

Building on its regional base, oil had some advantages in its competition with coal. Foremost was price. Oil was a bargain at less than $1.00 a barrel. Even when prices rose, the low cost of equipment conversion and shipping gave oil an advantage over the more cumbersome coal. Moreover, compared to the fragmented, labor-intensive coal industry, the well-financed and effectively managed major oil companies were efficient enterprises (Melosi 1985, pp. 35-50).

Energy Use and Pollution in the Industrial Age

Growing competition between coal and petroleum speaks to the vast energy abundance of the nation. Competing for markets took precedence over conserving resources; unrestrained economic growth was the credo of the day. The concept of "environmental cost," as a consequence of doing business, did not find its way onto the balance sheet of companies in the period.

The development and use of any energy source is environmentally intensive. The exploitation of wood led to deforestation and erosion. Mining of anthracite scarred northeastern Pennsylvania. The use of bituminous raised concern because of its dense, highly toxic smoke. From the 1870s through World War I, soft coal polluted the air of many urban areas, encroaching well beyond the city limits. Particularly hard hit were Pittsburgh, Cincinnati, St. Louis, and Chicago, where temperature inversions were common. The smoke problem was less significant in cities like New York, Boston, and Philadelphia which relied on anthracite (or San Francisco which used natural gas). If anthracite was unavailable, cities like New York turned to bituminous (Melosi 1985, p. 32).

Beyond the borders of industrial communities, beehive ovens for producing coke belched out hydrocarbons and waste heat, which killed foliage, trees and crops, leaving layers of dust and ash everywhere (Tarr n.d., pp. 6-7). Steam locomotives sprayed dense smoke and cinders all along their routes. Industrial cities were often railroad centers as well, aggravating an already critical problem of air pollution from stationary burners. Railroad coal use accounted for 20 to 50 percent of the smoke in Chicago and Pittsburgh (Tarr and Koons 1982, pp. 73, 77).

The debilitating effects of smoke on people became manifest by the 1890s. Methods of burning coal were so primitive that great amounts of heat, sulfur, and ash went up the smokestacks and chimneys. Newspapers reported chronic "Londoners" (a combination of smoke and fog) in several cities, which led to work stoppages, shortening of the school day, and many accidents. Although there were few scientific measures for smoke pollution, the assault on the senses set off protests. Citizens in Pittsburgh and St. Louis complained about frequent nasal, throat, and bronchial problems. Some observers speculated that deaths from pneumonia, diphtheria, typhoid, and tuberculosis could be traced to smoke, as could psychological trauma. Sooty walls of buildings, corroding marble statues, ash on hanging laundry, and grime on light-colored clothes were further testaments of the billowing black clouds (Grinder 1980, pp. 83-103).

By comparison, pollution in the oil industry rarely received the attention that smoke garnered. Part of the reason was that most of the oil fields were distant from urban areas. However, the euphoria over striking oil was not matched by restraint in producing and marketing it, resulting in substantial waste and serious spills caused by poor drilling and storage, natural disasters, the competitive market, simple disregard, and greed.

The problems oil men encountered across the continent were first experienced in Pennsylvania. While drilling practices steadily improved, storage of oil rarely was provided for until after a strike, and even then was usually inadequate. Earthen pits or wooden tanks were used, which resulted in substantial loss from evaporation or fire. Seepage from wells was common because of unreliable casings. In June 1892, Oil City experienced one of the worst disasters in petroleum history. A bursting dam on Oil Creek knocked over a huge tank of naphtha, blanketing the water and filling the air with flammable gas. A spark from a passing train ignited the fumes, engulfing tanks of crude in flames. About 300 people died in the conflagration.

Poorly constructed barrels carted over bumpy roads led to spillage, and barge accidents on waterways were frequent. Less obvious problems, such as water infiltration into oil strata, plagued the Pennsylvania fields. A newspaper noted in 1861, "So much oil is produced it is impossible to care for it, and thousands of barrels are running into the creek; the surface of the river is covered with oil for miles below Franklin" (Ise 1926, p. 25).

The pattern of waste and disregard for conservation were remarkably similar at Spindletop, Texas, where oil was struck in 1901, despite years of experience in drilling for oil in other locations. Fires periodically spread across the fields, such as one that burned 62 derricks and sent flames 1,000 feet into the air. Safety measures were eventually employed, but only after heavy losses. In 1902, the *Oil Investors' Journal* estimated that 10 million barrels had been wasted since the first strike. To impress investors, promoters sometimes opened up wells, sending gushers 125 feet into the air. The Spindletop field soon was ruined.

Overproduction and squandering of supplies were the result not only of shortsightedness; they were linked to prevailing competitive economic policy. Geologists in the 1870s warned about the dangers of extracting oil too rapidly, but few heeded the warnings. The "Rule of Capture," which dominated production until the 1930s, stated that those who owned the property over a common pool could take and keep as much oil as possible, regardless of the drainage from adjoining tracts. This fostered rampant drilling and pumping (Melosi 1985, pp. 47-49).

Drilling and refining created other environmental problems. Drain-offs of crude in the fields soaked the ground adjacent to wells. Rapid pumping introduced saltwater into the underground pools as well as into local water supplies. Floods along the coast washed oil into the rivers, lakes, and the Gulf of Mexico. In the pre-automobile days, oil even contributed to air pollution. Several days after the original well blew at Spindletop, a thick, yellow fog laden with sulfur engulfed Beaumont and did so periodically until production slowed. As Joseph Pratt claimed, "At least as much oil probably found its way into the region's ground water and air in this period as found its way to market" (Pratt 1978, p. 4). But while localized pollution in refining areas was serious, it rarely attracted much attention from oil companies or state governments before World War I. With immediate profits in mind and economic growth virtually a religion, pollution control was a luxury at best. As with coal, the equating of oil with progress precluded serious attempts to understand or address its various environmental impacts (Melosi 1985, pp. 49-50).

Factories as Polluters

The emergence of large-scale industry fueled by coal and oil reshaped the economy's infrastructure. The need for efficiency and economies of scale in the production phase led to the building of large factories. Because of their size, operational practices, and concentration near cities, modern factories--especially those of the textile, chemical, and iron and steel industries--became major polluters.

Factories were often located near watercourses, because water was needed for steam boilers or for other processes. Waterways also provided the least expensive means of disposing of soluble or suspendable wastes such as phenol, benzene, toluene, arsenic, and naphtha (Melosi 1988, pp. 753-61). In 1900, 40 percent of the pollution load on American rivers was industrial in origin. By 1968 that figure had increased to 80 percent (American Public Works Assoc. 1976, p. 410). The "death" of New Jersey's Passaic River in the late nineteenth century was a classic illustration of how factories defiled their surroundings. Before it became badly polluted, the Passaic was a major recreational area and the source of a thriving commercial fishery. As industrialization accelerated, the volume of sewage and waste pouring into the river forced Newark to abandon the Passaic as a water supply. Pollution also ruined commercial fishing, and soon homes along the waterway disappeared. During hot weather the river emitted such a stench that many factories were forced to close (Galishoff 1975, pp. 54-55).

The presence of a factory often meant the deterioration of the environs. Factories usually employed the simplest--not the most sanitary--disposal methods for garbage, slag, ashes, and scrap metals. Meat packing, which concentrated in cities such as Chicago and St. Louis, enveloped adjacent areas with foul smells and dumped animal wastes on vacant lots. Tanneries contributed more pollution by washing hides in available water sources (Galishoff n.d., pp. 4-5).

Factories also contributed an even more insidious, if not obvious, form of pollution--noise. Many factories produced high noise levels from machinery that was inadequately lubricated or not equipped with mufflers or arresters. High noise levels annoyed nearby residents, but also impaired the hearing of employees. In time, some businessmen recognized that factory din was unproductive and sought remedies, but most tolerated the noise since profit came from production, not conservation (Smilor 1977, pp. 28-29).

The Industrial City and Population Growth

Human concentration in major cities contributed notably to the pollution load of the era, especially when services lagged behind growth.

Between 1850 and 1920, world population increased 55 percent, but more than 350 percent in the United States, to 106 million. By 1920, 51 percent of all Americans lived in cities and the number of urban areas grew from under 400 in 1850 over to 2,700 in 1920 (Klein and Kantor 1976, pp. 69-71).

The rapid rise in population and urbanization was due primarily to dramatic increases in immigration and rural-to-urban migration. Almost 32 million people in these years came to the United States, largely from eastern and southern Europe. By 1910, 41 percent of urbanites were foreign-born, and approximately 80 percent of the immigrants settled in the Northeast. Migration from the countryside was at least 15 million between 1880-1920 (Dinnerstein and Reimers 1975, pp. 36-40; Klein and Kantor 1976, pp. 70-72).

Such population growth produced significant physical strain on most cities--and on the people who lived in them. None suffered more than the working class. Forced to live close to their places of employment, workers were crammed into spreading slums in central cities. Neighborhood densities were staggering. The thirty-two acres of New York City's Sanitary District "A" averaged nearly 1,000 people per acre in 1894--or approximately 30,000 people in a space of five or six blocks (Brody 1975, p. 133; Cochran and Miller 1961, p. 264).

Such crowded conditions and limited city services offered fertile ground for health and sanitation problems. In one of the most widely publicized epidemics of its day, Memphis lost almost 20 percent of its population in 1873 to yellow fever that allegedly started in the slums. In New Orleans, typhoid was spread by sewage oozing from the unpaved streets. In "Murder Bay" in Washington, D.C., mortality figures for black families were twice as high as in white neighborhoods. Many workers had little choice but to live in the least desirable sections of the city, usually close to the factories where they worked or near marshy bogs and stagnant pools. Environmental services often failed to keep up with demand. Smoke from wood-burning and coal-burning stoves and fireplaces fouled the air, and noise levels were deafening (Melosi 1980, pp. 10-12).

The industrial city rose skyward at its core as well as expanding outward into the hinterland, so not one but two distinct processes of growth contributed to its environmental problems. Pollution in the central city was most obvious, and initially more serious than in the suburbs. City governments were ill-prepared to provide necessary services. Refuse accumulated faster than it could be collected, and a relatively primitive system of wastewater removal was in effect until the turn of the century. Sanitation systems, when introduced, often served the business districts and only the better residential areas. In many cases, sewage systems, meant to eradicate effluents from the city proper, merely transferred the wastes to nearby rivers and lakes, shifting the problem downstream (Tarr and McMichael 1977).

Technological advances crucial to the development of the central business district also contributed to pollution. The balloon frame, the steel girder, and the elevator made possible high-density building, further straining meager services and added to congestion problems. Advances in transportation--the horsecar and electric streetcar--also intensified congestion and sometimes contributed directly to pollution--as in the case of horse manure. The wholesale uprooting of plants and trees reduced oxygen generation and transformed cities into "heat islands."

Outward expansion produced a variant to the environmental problems of the inner cities. Early suburbs suffered because they often were too far removed from needed services; later they suffered because they were too close to inner-city pollution (Melosi 1980, pp. 14-18).

The Search for Remedies

The confluence of air, water and land pollution, while not perceived as a crisis by contemporaries, nevertheless inspired a search for remedies in the late nineteenth and early twentieth centuries. Environmental problems usually were treated as localized occurrences. But sometimes, especially in the case of energy use and water pollution, the source of the problem was broader. The dilemma was how to reconcile economic benefits from industrial expansion with threats to the health and well-being of Americans and the degradation of the environment.

The lines of this dilemma were not as clearly drawn then as they are today. Few reliable tests existed to measure pollutants. In the field of public health, the bacteriological revolution in the 1880s offered the hope of eradicating epidemics. But "contagionists" and "anti-contagionists" squabbled over how to combat disease. Placing pollution in an ecological context was hampered by crude theory, and standards of environmental purity had yet to be established.

Despite these limitations and the significant economic achievements of industrialization, a search for remedies took several forms: civic protests, education programs, court action, regulation and legislation. With some exceptions, reform grew out of initiatives from environmental "consumers"--the receptors of the pollution or their agents (government)--rather than from "producers"--the generators of the pollution. But without cooperation from polluters or their acknowledgment that environmental costs were a consequence of doing business, early pollution abatement more often set precedents than it provided solutions.

Civic Protest

The earliest protests were responses to irritations, such as bad-tasting water, billowing smoke, putrefying garbage, or noisy machinery. The concept of pollution as a "nuisance" dominated these early complaints; indeed it pre-dates the industrial period. "Nuisance" was a popular contemporary term, applied arbitrarily to almost any environmental problem. "The great majority of the dwellers in our cities have not, heretofore, taken any active personal interest in the sanitary condition of their respective towns," said sanitarian John S. Billings. "They may grumble occasionally when some nuisance is forced on their notice, but as a rule, they look on the city as a sort of hotel, with the details of the management of which they have no desire to become acquainted" (Billings 1893, pp. 304-5).

By the 1890s, sporadic protests against the irritations of a dirty city led to individual and group efforts to deal more forthrightly with smoke, sewage, garbage, and noise. Reform groups pursued changes in nuisance laws and city ordinances, complained about industrial operations and municipal services, and criticized public behavior and conduct.

As more was learned about the generation of pathogenic organisms and disease transmittal, the concept of "health hazard" replaced "nuisance" as a way to describe an environmental problem. Health hazards encompassed communicable diseases such as cholera, yellow fever, typhoid, and dysentery, but also included afflictions such as respiratory diseases associated with smoke or emotional disorders attributed to high noise levels. Lack of medical knowledge produced fears of smells, miasmas, and water discoloration. But while reformers sometimes exaggerated a potential threat, they faulted on the side of caution.

By about 1900, the concern for health expanded into a broader environmental perspective, indicating that reformers were beginning to see pollution not simply as an irritant but as an unwanted by-product of industrialization. Pollution was sometimes linked to wastefulness and inefficiency, but in such a way as to avoid the conclusion that industrial activity was intrinsically responsible for despoiling the environment. In many ways, reformers attempted to find ways to mitigate the excesses of industrialization without abandoning its economic benefits.

Some viewed pollution in more abstract terms: cleanliness was a sign of civilization; pollution was barbarity. Excessive noise, E. L. Godkin (quoted in Smilor 1977, p. 26) asserted, "invades the house like a troop of savages on a raid, and respects neither age nor sex."

At least two strains of urban environmental reform emerged in the late-nineteenth century. One was composed of professionals with technical and scientific skills--sanitary engineers, medical officials, efficiency experts--who worked primarily within the municipal bureaucracy. They developed systems to combat health hazards and pollutants, compiled statistics, and monitored some forms of pollution. Sanitary engineers devised strategies for street cleaning and refuse disposal, designed sewers and drainage systems, and developed new methods of ventilating buildings. Efficiency experts installed cost-accounting systems and organized personnel-management programs. Health officials promoted environmental sanitation and epidemic control. As a group, they provided expertise never before available to municipal government.

Although the expert elite transmitted their ideas to municipal policy makers and through professional organizations, they were largely ineffective in communicating environmental concerns to the public. A second group filled the void. They were citizen organizations with strong civic and aesthetic values who operated outside city government, generating influence through protest or public awareness. Lacking the opportunity to implement changes themselves, they often supported efforts of the technical elite. The heart of this drive--often led by women--was voluntary citizens' associations, reform clubs, and civic groups whose interest in urban life was diverse. Sometimes environmental pressure groups, such as smoke- and noise-abatement leagues, formed to combat specific threats. Membership of the groups came from the middle- and upper-middle classes, and cut across several professions. Thus they were often insulated from many of the worst environmental threats. Their environmentalism was essentially a general concern for civic improvement (Melosi 1982, pp. 36-37).

Despite the transformation of intermittent civic protest into a more coherent form, pollution did not disappear. Public awareness increased, but protests needed to produce action. Faith that experts would find solutions was widespread. While that faith was overblown, technical and scientific methods in fact often made a difference. This was especially true in public health, where bacteriological laboratories monitored water supplies and successfully battled communicable diseases. Engineering expertise applied to implementing environmental services and improving efficiency of machinery also offered hope for reducing pollution.

The most severe restriction on reform in the early twentieth century was the lack of a broad environmental perspective. Pollution problems were most often addressed as isolated cases; little attention was paid to root causes. Interrelationships among various pollutants seemed to escape attention. Other than civic groups, no institutions were primarily concerned with environmental quality. From the consumer's vantage point, eliminating the pollution--not the source--was the goal (Melosi 1980, pp. 20-21).

Education

The hope that polluters would change their ways and that citizens could become more effective led to experiments in environmental education stressing behavioral change. In the area of smoke abatement, debate raged over the feasibility of education as opposed to the prosecution of violators--a debate, for example, that fractured the coalition of anti-smoke groups in St. Louis in 1911. In several quarters the call for education prevailed because supporters believed that promoting efficiency and fuel economy would stimulate *additional* economic growth (Grinder 1980, pp. 95-96). In Chicago, the Society for the Prevention of Smoke-- founded by local businessmen in 1892--initiated an educational campaign for the city's downtown business community to reduce an unwanted nuisance, to minimize a costly economic problem, and to improve the city's image. The goal was to achieve voluntary compliance through encouraging businessmen to employ new boiler designs and to retrofit existing boilers with smoke consuming devices (Rosen 1989, p. 6ff).

Similar programs were instituted in other cities to help eradicate various forms of pollution. New York's Society for the Suppression of Unnecessary Noise allied itself with the press in a campaign meant to inform the public about the tyranny of noise and to urge the writing of anti-noise ordinances. One result was the establishment of quiet zones around hospitals and schools (Smilor 1980, pp. 143-44).

But education of polluters alone, without more coercive measures, failed to produce many immediate or tangible results. A potentially more effective, if not more insidious, use of the education strategy was proselytizing among the young with the hope that changing behavior of children would have long-term benefits. A good example was the Juvenile Street Cleaning League, initiated by New York Street Cleaning Commissioner George E. Waring. Noting the participation of children in the local Civic History Club and other organizations, Waring concluded that "it seemed possible to enlist their interest in the cleanliness of the city." He hoped the children, especially the offspring of immigrants, would act as eyes, ears and noses for the department in uncovering unsanitary conditions, and would influence their parents. Like other middle-class reformers, Waring mistakenly viewed the working classes as the primary cause of indiscriminate waste disposal. In 1896 neighborhood leagues popped up throughout the city. There were 75 leagues and 5,000 participants in New York by 1899. Other cities, including Philadelphia, Brooklyn, Pittsburgh, Utica, and Denver, established their own leagues (Melosi 1981, pp. 74-77).

The Courts

Some contemporaries believed that self-regulation and the proper application of technology were practical remedies as long as all parties were aware of the nature and extent of pollution problems and their potential impacts on business efficiency and economy. Others placed faith in the coercive power of government to contain the crass impulses of would-be polluters. Still others assumed that the courts would apply traditional principles of common law to protect property rights and public safety.

Throughout the nineteenth century, interpretation of common law failed to meet the demand for preventive protection of the environment. But in the application of nuisance law, it developed a legal form for addressing the issue of liability for damage. Indeed, much of the legal history of the environment has been written by nuisance law. However, nuisance case law and statutes were not developed to address environmental issues, particularly before harm was done. Thus while no common law doctrine is wider in scope than nuisance law, its application to environmental issues has neither been systematic nor preventative (Watrous 1901, p. 98; Reynolds 1978; Krauss 1984, p. 250; Bone 1986).

In the nineteenth century, economic as opposed to ecological considerations shaped the application of nuisance law in cases affecting the environment. Lawrence Friedman and others have stated that judges deliberately structured tort doctrine to favor the interests of burgeoning industries. Morton Horwitz added that nineteenth century tort law provided an indirect economic subsidy to entrepreneurial interests, but that nuisance torts were slow to respond to the demand for economic subsidization (Horwitz 1977, pp. xi-xiv, 74).

Actions for injunctions against industrial nuisances increased after 1837. Between 1871 and 1916, the number of private suits for nuisance injunctions multiplied at a rate comparable to the remarkable economic growth of the period. But denial of injunctive relief persisted (Horwitz 1977, pp. 651-70).

The incidence of smoke pollution is a good example of the application of nuisance law in the wake of economic growth. Smoke was a visible sign of economic prosperity and material progress in those cities most dependent on activities causing it. Few courts were willing or able to determine the costs of smoke in hard economic terms, thus the relatively anemic legal response to air pollution at the time (Laitos 1975, pp. 66-70).

While the courts tended to protect entrepreneurs from nuisance actions, some progress was visible. In hearing claims for damages, even narrow interpretations of the law left room for successful later actions against industrial polluters (Schwartz 1981, p. 1775). In some states--New York, Pennsylvania and New Jersey--victims of pollution won relief much more frequently than they were denied it, especially after 1871 (Rosen n.d., pp. 8-11).

With respect to municipal and industrial wastes, a swing toward plaintiffs is not evident. Part of the reason is that: (1) the courts in the nineteenth century had no conclusive evidence that inorganic industrial waste posed a hazard to water supplies; (2) land disposal was treated as a relatively insignificant nuisance with little regard for its potential in contaminating groundwater; and (3) the introduction of municipal sewage into running water was not seen as a public nuisance. The courts continued to uphold riparian rights and dealt with nuisance cases as they arose rather than addressing them as part of a more generalized problem. The courts generally held polluters liable for damage caused by disposal in streams, but the legal system gave few prohibitions against the act of polluting itself (Fetig 1926; Besselievre 1924, pp. 217-19; Colten 1987, p. 8).

Government Regulation

During the period of rapid industrialization of the nineteenth century, state and federal governments, as well as the courts, usually took a position that encouraged economic growth and large-scale extraction of energy sources. But because the public domain held vast reserves of fossil fuels, federal officials began to realize the contradiction between promoting economic growth and providing stewardship over public lands. The "wise use" concept of resource conservation, emerging at the turn of the century, was a happy compromise (Melosi 1985, pp. 79-85).

A major link between conservation and energy production was the disposition of mineral resources on public lands. Debate over coal leasing during the Progressive Era involved questions of resource exploitation and control over the use of public lands (Hays 1959, pp. 82-90; Smith 1966, p. 115; Ise 1926, p. 324ff; Penick 1968, p. 77ff). Controversy over the federal role in waterpower also went to the heart of "wise use," an issue central to exploiting energy sources. By the early twentieth century, America's waterways came to be viewed as multipurpose resources, including the generation of hydroelectric power. But private companies had already gobbled up western waterpower sites on most public lands. Not until the Water Power Act (1920) did the federal government assert the principle of public regulation of hydroelectric power, including a clearer sense of waterpower development on navigable streams (Penick 1968, pp. 48-58; Hays 1959, pp. 74-81, 96-97, 160-65, 192-95). With respect to hydropower and the leasing of mineral lands, the issue was controlling access to and the allocation of resources on the public domain.

By World War I, the federal government began to restrain the unlimited exploitation of the public domain, but it did not broaden the "wise use" concept sufficiently to include those resources under private control. State officials in particular explored remedies for potential oil depletion, but wasteful practices continued. By the late 1920s the realities of a glutted market encouraged the oil industry to develop an appreciation for conservation. Discoveries in California and Oklahoma meant that known reserves of crude had greatly outstripped demand. The net result was incredibly low prices, a trend that worsened in the 1930s. Only after wanton exploitation of fields in east Texas and Oklahoma in the early 1930s was mandatory pro-rationing of well pumping enacted. At that point some order returned to the fields. State agencies such as the Texas Railroad Commission exercised greater influence over production controls than federal policies (Melosi 1987, p. 169).

If the broadening of the "wise use" concept for energy sources was slow in coming, plans for protecting the environment were glacial. The federal government's approach was sporadic and particularistic, reflecting the general perception that these were local problems best dealt with by ordinances or other local action. Congress passed few laws that dealt with environmental implications of energy. As late as 1955 government-sponsored environmental programs represented only 3 percent of the federal budget, with no financial support for pollution abatement (Rosenbaum 1977, p. 12).

Despite the lack of federal leadership, industrial cities paid some attention to the environmental implications of energy exploitation and use after the turn of the century. The greatest success of smoke abatement proponents was the implementation of tougher local laws in almost every city by 1912. Yet local authorities were unwilling to curb industrial development and selectively enforced the ordinances. During World War I, when unrestricted production became a patriotic duty, smoke abatement fell on hard times. Smoke pollution did not subside until the use of coal diminished in the 1920s (Grinder 1980, pp. 83-103; Tarr and Koons 1982, pp. 71-92).

Pollution from oil production and transportation also remained mostly a local issue. Oil drilling and refining polluted local areas but rarely attracted attention from oil companies or state governments. The federal government, however, established a precedent when it passed the Oil Pollution Control Act in 1924. Contamination of water from tanker discharges and seepage problems on land were the primary problems. The former attracted most attention largely because the polluting of waterways and coastal areas affected commercial fishermen and resort owners. The oil industry at first treated the call to end polluting practices with apprehension. But implementation of the law, and progress towards halting the wide-spread pollution by oil, were slowed when the American Petroleum Institute, the industry's major lobbying body, realized that it could control the flow of information to the government and hence influence pollution control efforts.

Herbert Hoover was the leading government proponent of oil conservation and of curbing pollution during the 1920s. As secretary of commerce he tried to protect American fisheries, despite his additional responsibility to commercial shippers. Hoover and his supporters wanted a comprehensive law to regulate land-based polluters as well as ships. But conflicting economic and political interests in Congress produced a much weaker bill. The 1924 Act had inadequate enforcement provisions and dealt only with dumping fuel at sea by oil-burning vessels. Although the act disappointed Hoover and the conservationists, it was the first serious attempt to deal with the issue on a national scale (Melosi 1987, pp. 170-71).

While the environmental implications of energy use and development received sporadic attention at the local, state, and federal levels before World War II, nineteenth-century American municipalities had written an array of ordinances regulating activities associated with health and sanitation, including the construction and emptying of privy vaults and cesspools, garbage collection, sewerage development, and water supply.

By the 1880s many cities passed statutes forcing "noxious" manufacturers to the outskirts, but few cities had specific regulations regarding disposal of manufacturing wastes. If they were dealt with at all, these wastes were lumped under existing nuisance provisions (Tarr 1985, p. 96; Colten and Breen 1986, pp. 54-55).

But municipal statutes and ordinances did little to move beyond existing judicial definitions of environmental liability. Municipalities sometimes found themselves as defendants, especially when the dumping of sewage threatened a riparian owner. In addition, the diversion of industrial wastes to municipal treatment facilities transferred legal responsibility for stream pollution to the cities (Tooke 1900; Fertig 1926, p. 786; Colten 1987, p. 10).

While municipal ordinances infrequently confronted the broad problems associated with land and water pollution, establishment of regulatory bodies in some cities worked to decrease pollution. The Chicago Drainage Districts attempted to prevent more pollution in already badly contaminated watercourses. The Milwaukee Sewerage Commission was empowered to regulate the character and quantity of industrial waste discharged into public sewer systems (Jackson 1924, p. 23; Jackson 1937, p. 86).

The states were the centers of action for new legislation to control stream pollution--which remained the major focus of concern over the disposal of municipal and industrial wastes through the middle twentieth century. In general, the number and scope of sanitation laws increased dramatically on the state level by the end of the century--particularly in the Northeast and less so in the South. The first state legislation to control stream pollution was written in 1878 in Massachusetts. It gave the State Board of Health the power to control river pollution caused by industrial wastes. In 1915 only eighteen state boards of health had divisions of sanitary engineering; all but four states established such divisions by 1927 (Vesilind 1981, p. 26; Micklin 1970, p. 131). By World War I, states established boards and commissions expressly designed to regulate water pollution. These new bodies often were given expanded power over industrial as well as municipal pollution (Warrick 1933, p. 496; Monger 1926, p. 790; Tobey 1939, p. 1322).

Results were often disappointing, however. Regulations sometimes conflicted and enforcement was lax. A survey conducted by the American Water Works Association in 1921 stated that only five states granted ample authority to its pollution agencies, and in nearly all cases enforcement was hampered by lack of appropriations. Some experts believed that conditions improved by the late 1920s, citing the Ohio State Department of Health as an example of an effective regulator. But the early laws made substantial accommodations to industry. Some laws failed to provide penalties for infringements, and most laws exempted specific industries, specific streams, or specific wastes, such as petroleum, wood wastes, acids, and alkalis (Donaldson 1921, p. 198; Besselievre 1924; Fales 1928; Skinner 1939, p. 1332).

Under the terms of state legislation the individual or company responsible for a nuisance, in theory, could be subject to criminal action for violating a law prohibiting pollution by industrial wastes. In general, state boards preferred cooperation to placing themselves in an adversarial relationship with industry. State agencies often justified cooperation on the grounds that drastic control by court action would hinder economic growth and might result in incomplete investigations of actual stream requirements and of the applicability of treatment processes. Among the boards' functions was providing industries with technical information and survey data to keep them abreast of current disposal and recycling methods (Baity 1939, pp. 1302-03; Rue 1929; Besselievre 1924; Fertig 1926, p. 786; Tobey 1939, p. 1322). Trade associations, such as the American Petroleum Institute and the Pulp and Paper Association, were involved in projects for pollution control or waste utilization, although their motives could be suspect.

Another important wrinkle in the evolution of environmental liability at the state level was interstate conflict. On several occasions one states brought action against another state (or industry in that state) to restrain air and water pollution (Lay 1931; Tobey 1925). It did not take long to recognize that such chronic interstate rivalry could spill over into other areas. One possible solution was interstate agreements or compacts to control or abate pollution (Control of pollution 1939). The interstate cooperative approach had practical and salutary benefits for dealing with stream pollution, but it was an incomplete tool. The interstate compacts did not serve as regional plans to abate pollution, since they were drawn only to deal with the level of discharge into water. They also did little to further the definition of environmental liability, working instead to solve practical problems by avoiding litigation.

To a limited degree, federal regulation in the period sought to accomplish--for public health and water pollution at least--what state and court actions could not. There was, however, no overriding national policy for dealing with pollution. In 1912, the federal government began to give assistance to the states in evaluating water pollution through the Public Health Service's Stream Investigation Station in Cincinnati. In 1938 a loan and grant program for the states was set up through the PHS's

Division of Water Pollution Control. Some concern about hazardous substances was noted in several acts, especially the Food, Drug and Cosmetic Act of 1938, which limited the amount of additives and residues in agricultural products. Industrial safety and hygiene also received a national hearing by mid-century (Davies 1970, pp. 38-40; Baity 1939, pp. 1300-06; Tobey 1926).

Two pieces of legislation, although modest in the short-term, became important precedents for future action in dealing with water pollution and industrial waste: the Oil Pollution Control Act (1924), and the Refuse Act--the popular name for Section 13 of the Rivers and Harbors Act (1899)--prohibited the discharge of wastes (other than sewer liquids) into navigable waters without a permit from the Corps of Engineers. Violations were punishable by fine or imprisonment. This provision superseded the Refuse Act of 1890 which prohibited dumping that would "impede or obstruct navigation."

The 1899 act suggested a strict prohibition against dumping that seemed to go beyond the primary goal of the law, that is, to preserve waterways for navigation. In 1910, a New York group tried to invoke the act against a proposed sewer, but the Judge Advocate General ruled that pollution control was a function of the states alone. While one court in 1918 interpreted the act to forbid dumping *per se*, most courts interpreted the act literally in the early twentieth century. By the 1960s it was used, as one commentator noted, as a "*cause celebre* for the environmental movement." Or as another suggested,

> . . . a piece of legislation that was aimed at keeping carcasses of cows and other floating debris from obstructing the smooth flow of commerce seems to have been turned into a useful bit of antipollution legislation by some enterprising conservationists and politicians concerned with the environment.

In many ways, the Refuse Act became a complement to other federal water-quality legislation and a predecessor to the Water Quality Act of 1965 (Cowdrey 1975; Rodgers 1971a, 1971b).

Conclusion

The transformation of the American environment in the era of the fossil fuels began without serious attention to environmental costs as a consequence of doing business. Until the early twentieth century, environmental costs--if they were recognized at all--were missing from the economic balance sheet, were a relatively low priority in the political arena, and were inconclusively dealt with in the courts. Pollution was regarded as an aberration rather than as a predictable by-product of environmentally intensive economic activity. When pollution was not ignored or dismissed, it was considered an inevitable consequence of economic growth--rendering producers free of responsibility for its generation.

Political solutions to environmental problems required broadscale acceptance of environmental costs as an integral part of the production process and as an unacceptable side effect of economic growth. But most politicians were no more receptive to promoting long-term goals such as environmental protection than most businessmen were willing to sacrifice short-term profits.

The national commitment to economic growth, the dominance of big business, and the limited power--and resolve--of government offered few opportunities to confront the problem of pollution in any serious way. Ironically, squandering of natural resources, inefficient burning of fuels, and the tainting of watercourses also worked against what was perceived as the long-range economic interests of the country.

In essence, the Industrial Revolution of the nineteenth century produced a fundamental dilemma in the twentieth: Should an economic future, consistently shaped by a series of short-term goals, be revised to embrace a less finite and possibly less certain outcome shaped by attention to environmental costs? For the most part, the answer--in action if not in word--has been "no."

Acknowledgments

I would like to thank the DeLange Lecture Series at Rice University, the National Endowment for the Humanities, and the Energy Laboratory at the University of Houston for their financial support in preparation of this piece.

References

American Public Works Assoc. 1976. *History of Public Works in the United States, 1776-1976.* Chicago.

Baity, H. G. 1939. Aspects of governmental policy on stream pollution abatement. *American J. Public Health* 29, 1297-1307.

Besselievre, E. B. 1924. Statutory regulation of stream pollution and the common Law. *Trans. American Inst. Chem. Engineers* 16, 217-30.

_____. 1952. *Industrial Waste Treatment.* New York: McGraw-Hill.

Billings, J. S. 1893. Municipal sanitation: Defects in American cities. *The Forum* 15, 304-310.

Binder, F. M. 1958. Anthracite enters the American home. *Pennsylvania Magazine of History and Biography* 82, 82-99.

Bone, R. G. 1986. Normative theory and legal doctrine in American nuisance law: 1850-1920. *Southern California Law Review* 59, 1142-1226.

Brody, D. 1975. Slavic immigrants in the steel mills. In *The Private Side of American History*, ed. Thomas R. Frazier. New York: Harcourt, Brace Jovanovich.

Burlingame, R. 1946. *March of the Iron Men.* New York: Charles Scribner's Sons.

Chandler, A. D., Jr. 1972. Anthracite coal and the beginnings of the industrial revolution in the United States. *Business History Review* 46, 141-81.

Cochran, T. C., and W. Miller. 1961. *The Age of Enterprise.* New York: Harper Torchbooks.

Cole, A. H. 1970. The mystery of fuel wood marketing in the United States. *Business History Review* 44, 339-59.

Colten, C., and G. Breen. 1986. *Historical Industrial Waste Disposal Practices in Winnebago County, Illinois, 1870-1980.* Champaign, Ill.: Illinois Hazardous Waste Resources and Information Center.

Colten, Craig E. 1987. Industrial wastes before 1940. Unpublished paper.

Control of pollution in interstate waters. 1939. *American City* 54, 58-60.

Cowdrey, A. E. 1975. Pioneering environmental law: The Army Corps of Engineers and the Refuse Act. *Pacific Historical Review* 44, 331-49.

Davies, J. C., III. 1970. *The Politics of Pollution.* New York: Pegasus.

Degler, C. N. 1967. *The Age of the Economic Revolution, 1876-1900.* Glenview, Ill.: Scott, Foresman.

Dinnerstein, L., and D. M. Reimers. 1975. *Ethnic Americans.* New York: Harper and Roe.

Donaldson, W. 1921. Industrial wastes in relation to water supplies. *American J. Public Health* 11, 193-98.

Fales, A. L. 1928. Progress in the control of pollution by industrial wastes. *American J. Public Health* 18, 715-27.

Fertig, J. H. 1926. Legal aspects of the stream pollution problem. *J. American Public Health Assoc.* 16, 782-88.

Galishoff, S. 1975. *Safeguarding the Public Health: Newark, 1895-1918.* Westport, Conn.: Greenwood.

_____. n.d. Sanitation in nineteenth and early twentieth century urban America: An overview. Unpublished paper.

Grinder, R. D. 1980. The battle for clean air: The smoke problem in post-Civil War America, 1880-1917. In M. V. Melosi. ed. *Pollution and Reform in American Cities, 1870-1930.* Austin: University of Texas Press.

Hays, S. P. 1959. *Conservation and the Gospel of Efficiency.* New York: Atheneum.

Hindle, B. 1975. *America's Wooden Age.* Tarrytown, New York: Sleepy Hollow Press.

Horwitz, M. J. 1977. *The Transformation of American Law, 1780-1860.* Cambridge, Mass.: Harvard University Press.

Hunter, L. 1979. *Waterpower in the Century of the Steam Engine.* Charlottesville: University Press at Virginia.

Ise, J. 1926. *The United States Oil Policy.* New Haven: Yale University Press.

Jackson, J. F. . 1924. Stream pollution by industrial wastes, and its control. *American City* 31, 23-26.

_____. 1937. Industrial wastes in city sewers-I. *American City* 52, 86-87.

Kirkland, E. C. 1969. *A History of American Economic Life.* New York: Appleton-Century-Crofts.

Klein, M., and H. A. Kantor. 1976. *Prisoners of Progress.* New York: Macmillan.

Krauss, E. P. 1984. The legal form of liberalism: A study of riparian and nuisance law in nineteenth century Ohio. *Akron Law Review* 18, 223-53.

Laitos, J. G. 1975. Continuities from the past affecting resource use and conservation patterns. *Oklahoma Law Review* 28, 60-96.

Lay, G. C. 1931. Suits by states to abate nuisances. *United States Law Review* 65, 73-85.

Melosi, M. V. . 1980. Environmental crisis in the city: The relationship between industrialization and urban pollution. In *Pollution and Reform in American Cities,* ed. M. V. Melosi. Austin: University of Texas Press.

_____. 1981. *Garbage in the Cities.* College Station, Texas: Texas A&M Press.

_____. 1982. Battling pollution in the Progressive Era. *Landscape* 26, 35-41.

_____. 1985. *Coping with Abundance.* New York: Knopf.

_____. 1987. Energy and environment in the United States: The era of fossil fuels. *Environmental Rev.* 11, 167-88.

_____. 1988. Hazardous waste and environmental liability: An historical perspective. *Houston Law Rev.* 25, 741-79.

Micklin, P. P. 1970. Water quality: A question of standards. In *Congress and the Environment,* ed. R. A. Cooley and G. Wandesforde-Smith. Seattle: University of Washington Press.

Monger, J. E. 1926. Administrative phases of stream pollution control. *Journal of the American Public Health Association* 16, 788-94.

Penick, J., Jr. 1968. *Progressive Politics and Conservation.* Chicago: University of Chicago Press.

Petulla, J. M. 1977. *American Environmental History.* San Francisco: Boyd and Fraser.

Powell, H. B. 1980. The Pennsylvania anthracite industry, 1769-1976. *Pennsylvania History* 47, 3-28.

Pratt, J. A. 1978. Growth or a clean environment? *Business History Review* 52, 1-29.

Reynolds, O. M., Jr. 1978. Public nuisance: A crime in tort law. *Oklahoma Law Review* 31, 318-43.

Rodgers, W. H., Jr. 1971a. The Refuse Act of 1899: Its scope and role in control of water pollution. *Ecology Law Quarterly* 1, 173-202.

———. 1971b. Industrial water pollution and the Refuse Act: A second chance for water quality. *University of Pennsylvania Law Review* 119, 322-35.

Rosen, C. M. 1989. Chicago's Society for the Prevention of Smoke: Education and the law in the fight against air pollution in the 1890s. Unpublished paper.

———. n.d. A litigious approach to pollution regulation: 1840-1906. Unpublished paper.

Rosenbaum, W. A. 1977. *The Politics of Environmental Concern.* New York: Praeger.

Rue, J. D. 1929. Disposal of industrial wastes. *Sewage Works Journal* 1, 365-69.

Russel, R. R. 1964. *A History of the American Economic System.* New York: Appleton-Century-Crofts.

Schurr, S. H., and B. C. Netschert. 1960. *Energy in the American Economy, 1850-1975.* Baltimore: Johns Hopkins University Press.

Schwartz, G. T. 1981. Tort law and the economy in nineteenth-century America: A reinterpretation. *Yale Law Journal* 90, 1717-75.

Skinner, H. J. 1939. Waste problems in the pulp and paper industry. *Industrial Engineering Chemistry* 31, 1331-35.

Smilor, R. W. 1977. Cacophony at 34th and 6th: The noise problem in America, 1900-1930. *American Studies* 28, 23-38.

———. 1980. Toward an environmental perspective: The anti-noise campaign, 1893-1932. In *Pollution and Reform in American Cities,* ed. M. V. Melosi. Austin: University of Texas Press.

Smith, F. E. 1966. *The Politics of Conservation.* New York: Harper Colophon.

Tarr, J. A. 1985. Historical perspectives on hazardous wastes in the United States. *Waste Management and Research* 3, 95-102.

———, n.d. Searching for a 'sink' for an industrial waste: Coke production and the environment. Unpublished paper.

Tarr, J. A., and K. E. Koons. 1982. Railroad smoke control: The regulation of a mobile pollution source. In *Energy and Transport,* ed. G. H. Daniels and M. H. Rose. Beverly Hills, Calif.: Sage.

Tarr, J. A., and Francis Clay McMichael. 1977. Decisions about wastewater technology, 1850-1932. *Journal of Water Resource Planning and Management Division, ASCE 103*, 47-61.

Taylor, G. R. 1951. *The Transportation Revolution, 1815-1860.* New York: Rinehart.

Tobey, J. 1925. Public health and the United States Supreme Court. *American Bar Association Journal* 11, 707-710.

———. 1926. Federal control of hazardous substances. *American Journal of Public Health* 16, 244-49.

———. 1939. Legal aspects of the industrial wastes problem. *Industrial Engineering Chemistry* 31, 1320-22.

Tooke, C. W. 1900. Pollution of running streams by sewage. *Municipal Engineering* 19, 87-90.

Vesilind, P. A. 1981. Hazardous waste: Historical and ethical perspectives. In *Hazardous Waste Management*, ed. J. J. Peirce and P. A. Vesilind. Ann Arbor, Mich.: Ann Arbor Science Pubs., Inc.

Warrick, L. F. 1933. Relative importance of industrial wastes in stream pollution. *Civil Engineering* 3, 495-98.

Wattrous, G. D. 1901. Torts, 1701-1901. In Faculty, Yale Law School, *Two Centuries' Growth of American Law, 1710-1901.* New York: Charles Scribner's Sons.

6

Exhaustibility of British Coal in Long-Run Perspective

G. N. von Tunzelmann

In 1865, the economist William Stanley Jevons--known today as the first British neoclassical economist--published his first major work, *The Coal Question*. The analysis Jevons made of coal supply and demand was in fact classical political economy. He relied on the Malthus-Ricardo theories of rent for the supply side analysis and Malthusian population theory for the demand side. He then applied the principle of diminishing returns from classical rent theory to British coal mining. He made two fundamental arguments. First, coal was an exhaustible resource, whose extraction costs would inevitably rise as mining continued to expand, and second, the rise in extraction costs and the inevitable depletion of the coal resource would undermine the competitive edge of the British manufacturing economy.

This chapter concentrates primarily on whether Jevons was right that coal is an exhaustible resource and, by implication, the larger question of exhaustibility of energy resources so important to human activities and so central to issues of the global environment.

Though Jevons's book was not the first to consider the exhaustibility of fuel ("the coal question"), it was nevertheless a landmark study because it took the subject from professional geologists and working coal miners and brought it to the British public at large. The book's immediate political impact was powerful: Gladstone, as Chancellor of the Exchequer, referred to it in his 1865 budget speech, and a Royal Commission was formed in June 1866 to address the coal question. The Commission report in 1871 referred explicitly to Jevons's work on its first page, but it came, as we shall see, to very different conclusions about imminent exhaustibility.

One of the important points Jevons made was the distinction between literal, physical depletion of coal and the operation of economic forces that would make coal too expensive to mine. Jevons wrote on this issue as follows:

> The expression 'exhaustion of our coal mines' states the subject in the briefest form, but is sure to convey erroneous notions to those who do not reflect upon. . . the gradual deepening of our coal mines and the increased price of fuel. Many persons perhaps entertain a vague notion that some day our coal seams will be found emptied to the bottom, and swept clean like a coal-cellar. . . . It is almost needless to say, however, that our mines are literally inexhaustible. We cannot get to the bottom of them: and though we may some day have to pay dear for fuel, it will never be positively wanting" (W. S. Jevons 1906).

A 1925 government analysis of the coal question compares coal mining with agriculture and expresses the point about diminishing returns poetically:

> The industry may indeed be imagined as not unlike a series of farms in a country of valleys and mountains, varying in their productivity from the fat lands by the rivers, through medium lands on the lower slopes, up, through farms of gradually decreasing fertility, to fields that are half rock at the limit of cultivation on the higher slopes. . . . The question for the agriculturalists [and thus miners] is how far up the mountain-side it is worth while to spend labour (Samuel Commission 1926, p. 45).

The progression to less accessible and less productive seams of coal would bring about steadily rising costs. Eventually the cost of getting coal out of the ground would raise the price to such a level that for most buyers it would be as if there were no coal at all.

But why did Jevons assert that "our mines are literally inexhaustible" when his larger argument was that extraction costs would price coal out of the market eventually? The argument he presented differed from the crude model of popular imagination, unfairly credited to Malthus, in which population growth would increase demand for coal until supplies were actually exhausted. At the growth rates of the mid-nineteenth century, Jevons showed that British coal reserves would run out in the 1970s under this simple model. But he did not believe (nor did Malthus before him) that events would unfold in this simplistic manner. Instead Jevons relied on Malthus's notion that population growth and food supplies (or living standards) would interact. In moderately advanced societies such as the Britain of his time, the interaction would curb food consumption (or lower living standards) by raising the prices of goods (von Tunzelmann 1986a). Population would not be permitted to grow at biologically determined rates, but would be checked, most simply through increased mortality (the "positive check"), or in more advanced societies through reduced fertility (the "preventive check").

The implication of such a model, applied to the coal question, was that demand for coal could not grow unchecked because the effect of diminishing returns in coal supply, confronting such a growth of demand, would be to drive up costs and prices of coal and eventually restrict such demand growth. Jevons indeed argued that population growth rates were already declining in Britain at the time he was writing, though of course he was more concerned about the check to continued industrial expansion from rising fuel costs. Hence coal resources could be regarded as "literally inexhaustible" because cost increases would choke off demand at a point well before the geophysical limit of extraction.

An additional complexity needs to be remembered. From the early seventeenth century in Britain (and from the mid-nineteenth century in the United States), coal substituted for wood as fuel. Clearly coal is a non-renewable resource; on human time scales coal cannot be created anew. However, while timber can be replanted and restocked within a couple of decades, expansion of timberlands involves land. In land-hungry nineteenth-century Britain, timber competed with humans for land for living space. Land was also non-renewable, other than to a very limited extent through reclamation. Thus, considered in this light, the separation of resources into exhaustible and renewable should be treated more critically than is normally the case. This recommendation applies also to resources such as water today (see the paper by Gleick in this volume).

Jevons's analysis considered the demand side and the supply side, followed by a consideration of their interaction. In this respect it differs from the orthodox economic theory of exhaustible resources, which derives from a seminal paper by Hotelling (1931) (see also Dasgupta and Heal 1979). Hotelling approached the issue of exhaustible resources as a capital budgeting problem. He showed that in order to encourage wealth holders to hold their wealth in the form of exhaustible resources, the price of the asset would have to rise over time. With perfect foresight and the standard range of assumptions of marginalist economics, such as perfect competition, the price would rise at a rate dictated by the rate of interest (effectively the sum of extraction costs plus the compound interest factor). One can show that, as a result of compounded interest, the major part of the increase in price would fall in the late years immediately preceding exhaustion (see, for example, Peirce 1989). The analysis can be modified to take account of the possibility that not all resources in a particular field, such as energy, are exhaustible. This modification was described by Nordhaus (1973 and 1979), who showed that, if a "backstop" renewable technology existed that would become available at some future date at a (fixed) price, then that would truncate the Hotelling compound-interest process. The analysis can also be modified to take theoretical account of such complications as imperfect competition. These modifications became essential to any attempt at studying such phenomena as OPEC and the oil crises of the 1970s (Dasgupta and Heal 1979).

It is apparent that changes in extraction costs as time passes are not immediately relevant to Hotelling's model, even though those changes have constituted the bulk of the empirical interest of economists, politicians and others in this question over the past century or more. Hotelling himself considered extraction cost changes in his original 1931 paper (see also Brooks 1974). The relative neglect of this matter by economic theorists must be ascribed to the static nature of equilibrium analysis in conventional neoclassical economics (von Tunzelmann 1990). A possible interpretation is that classical diminishing returns, represented by rising extraction costs, absorb the compound-interest factor assumed in conventional analysis to accrue to wealth holders. The conflict between the analyses then reduces to a question of political economy: which social class or classes benefit from the rising prices of the exhaustible resource. This question is obscured by neoclassical economics (Fine 1990).

That an exhaustible resource would follow over historical time a cost pattern such as Hotelling described seems intuitively reasonable; an empirical basis for this view is provided in the next section. The practical problem is that the evidence adduced to date has not supported Hotelling's model. The seminal study is by Barnett and Morse (1963), which claimed to demonstrate that, in the United States from 1870 to 1957, price and cost data did not indicate "increasing scarcity." Real prices of minerals, including coal, declined during the period Barnett and Morse considered, though the decline was concentrated in the first decade they examined--the 1870s. (Real prices are prices relative to an index of all output prices.) In contrast, real agricultural prices showed a fairly constant level over time, while forestry and fishing both showed some tendency to rise.

Julian Simon reopened the issue in 1981 (Simon 1981). He argued that the price of oil was the one possible exception to an otherwise general decline in real prices of resources over the past two centuries. (The trend in oil prices in the 1980s might eliminate even that exception.) Simon received considerable publicity for a wager, on behalf of the "anti-Malthusians," that real prices of resources would fall rather than rise during the 1980s; recently the "Malthusians," represented by Paul Ehrlich, conceded defeat.

In fact, however, Simon's position is not necessarily inconsistent with the more complex models Malthus used. Indeed, Malthus himself concluded that, in the five centuries up to the time of his writing, real agricultural prices had remained roughly constant in the long run, while real industrial prices had substantially fallen (von Tunzelmann 1991). Malthus drew such conclusions from applying to such secular trends the "labor-commanded" theory of value, which he aimed to advocate on theoretical grounds. Under that theory, price changes over time are compared to wages rather than an index of all output prices. What is especially paradoxical is that Simon, protesting his opposition to Malthusianism, advocates precisely the same type of measure of value over the long term. Simon's argument is that comparing the price of a particular good with other output prices is not an accurate indication of that good's scarcity or abundance because technical progress in the production of other goods can artificially raise the price of the good under consideration. It is as if the goalposts are being shifted in such a way as to make it more and more difficult for a team to score, though it is playing well. In economic terms, the price of a good rises because its productivity performance is below average, but still positive. To Simon--and Malthus-- a comparison of prices with wage rates provides a better indicator of absolute as opposed to relative changes in the productivity of natural resources.

This chapter reassesses these questions as applied to coal by posing them over an even longer historical framework than Barnett and Morse employed. Coal was chosen for the focus because it has been the basis of Britain's industrial economy for centuries--certainly since the Industrial Revolution--and is near the center of several important environmental problems arising from human activities. Our main objective is to assess the empirical foundation for the exhaustibility of resources, judged over the very long run of centuries.

At any particular time, the costs of producing coal will differ at various mines. Coal from some very efficient mines will cost less than average, many medium-efficiency mines will produce at average costs, and a few relatively inefficient mines will produce only at high costs. One should expect that the bulk of the coal would be produced at low and medium costs, and that only special circumstances would result in the sale of a relatively small amount of high-priced coal.

Figures 6.1, 6.2, and 6.3 provide three empirical views of the relationship between quantity of coal produced and its cost or price. Figure 6.1, prepared with data from 1981-82, matches the expected relationship quite well. It shows that in those years about 90 million tons of coal a year could be produced in Britain from efficient or moderately efficient coalpits with costs per ton not exceeding 50 pounds (von Tunzelmann 1986b and 1989). The upward-trending "tail" on the right of the graph indicates that somewhat less than 20 million tons came from higher cost mines.

Comparable figures for earlier years are only occasionally available. Figure 6.2 is based on figures for the whole country's pits in the last quarter of 1918, when the government was running the coal industry. The relationship between price and output looks surprisingly like that for the early 1980s, although the proportion of coal coming from very high-cost pits is smaller.

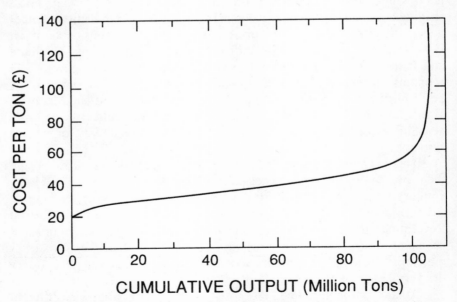

FIGURE 6.1 Quantity of coal mined annually in relation to cost per ton, 1981-82.

For the bulk of the nineteenth century, or earlier, no general cost survey is available, and too few data from individual mines exist to draw strong conclusions (but see Church et al. 1986 and Flinn 1984 for heroic attempts). One is compelled to use price data to approximate costs. The problem with so doing is that prices are dominated by demand influences and diverge from costs. It appears from the data displayed in the figures that no coal is sold at high prices, though we know that some coal is produced at high costs. The lack of homogeneity of coal as a product is an additional problem in trying to develop price series for coal. Figure 6.3 is based on 1837 prices in the North-East coal field, then Britain's largest coal-producing region. It suggests a pattern roughly similar to that of a century later, once we realize that the reconstructed data eliminate the "tail" shown in Figures 6.1 and 6.2.

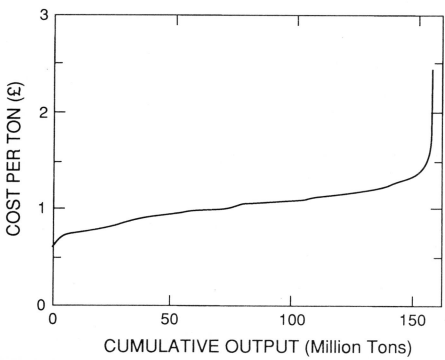

FIGURE 6.2 Quantity of coal mined annually relative to unit cost in 1918.

The conclusion to be derived from Figures 6.1, 6.2, and 6.3 is that coal costs at any point in time are likely to be distributed across mines in roughly the fashion indicated by the curves in those figures. The fact that coal is by no means a homogeneous product complicates the issue somewhat, but there is reasonable evidence for the existence of static diminishing returns, indicated by the general slopes of the curves and especially the right-hand tails.

A logical inference is that the extension of mining through time would lead further and further into the higher-cost regions of the curves, as the high-production, low-cost pits near exhaustion. Jevons applied the Ricardian model of diminishing returns to argue that, barring accidents of discovery--which were becoming less frequent in Britain by the nineteenth century--the most accessible and richest seams of coal would be the first to be exploited. As such shallow and thick coal seams became gradually exhausted, miners would move on. But they would necessarily have to open up deeper or thinner seams of coal, which require higher costs both for access and for working.

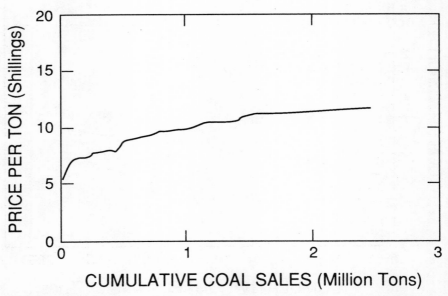

FIGURE 6.3 Price per ton of coal sold in the North East coal field in 1837.

Data on working costs are available from 1868, though not for every year. When deflated by an output price index--not by labor costs--as in Figure 6.4, they appear to show the effect that Jevons forecast: an exponentially rising cost strongly indicating diminishing returns over time. Real operating costs were about three times as high in the 1970s and 1980s as in the late nineteenth century, when Jevons was writing. It is also apparent that short-run movements reflect factors other than diminishing returns. In particular, the steep rise in real costs during the 1970s mirrors the oil-led energy crises of that decade.

Whether British coal mining has faced diminishing returns can be put in longer-term perspective by using price rather than cost data. Price data compare better over time than over space at a particular time because, over long periods of time, competitive forces are likely to maintain profits at "normal" levels. The main problem is adjusting for heterogeneity of coal market prices stemming from differences in product quality and other characteristics such as labor costs. The data beginning in 1700, shown in Figure 6.5, reflect "pithead" coal prices, in order to approximate as closely as possible the actual costs of mining. The general pattern of operating costs as in Figure 6.4 reemerges in the price data of Figure 6.5, with a generally rising curve from about the time of Jevons's study.

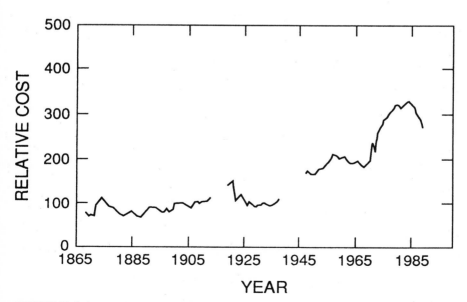

FIGURE 6.4 Relative cost of mining coal 1868-1989, deflated by output price index (1900=100).

A more surprising conclusion to be drawn from Figure 6.5 is that, in the nearly 200 years up to the time of Jevons writing, there was no perceptible influence of diminishing returns on real coal prices. Jevons himself was aware of this. He wrote that:

> . . . it appears that there has been no recent [cost] rise of importance [indicating exhaustion], but that, at the same time, the high price demanded for coals drawn from some of the deepest pits indicates the high price that must in time be demanded for even ordinary coals (W. S. Jevons 1906, 8 and 80).

Thus, for all the enormous expansion of demand for coal that went with the Industrial Revolution, there appears to have been no substantial effect on its real cost.

The data in Figures 6.4 and 6.5 were deflated by output price. If instead we pursue Julian Simon's suggestion and deflate production costs or prices by the price of labor, a very different picture emerges. This different picture is shown in Figure 6.6. It shows that real costs in terms of labor remained stable in the eighteenth century, declined in the middle years of the nineteenth, and, surprisingly, showed no trend increase in the

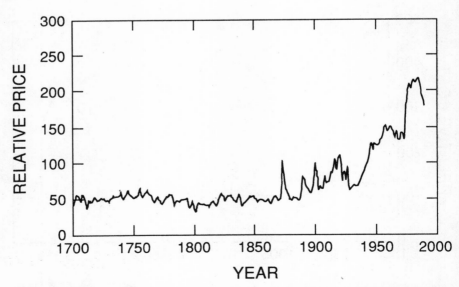

FIGURE 6.5 Relative supply price of coal at the pithead, 1700-1989, deflated by output prices (1900=100).

twentieth. The exponential cost growth of Figures 6.4 and 6.5 seems to vanish. Following Simon, it could be argued that twentieth century rising real costs deflated by all output prices are the result not so much of diminishing returns in the production of coal (or other exhaustible resources) as long-run "increasing returns" in the production of other goods--the consequences of technical progress in manufacturing more generally. Certainly this argument is plausible, but it fails to explain why coal technology could not match the advances in manufacturing, and thus does not answer the questions about diminishing returns in coal production.

The two kinds of indices of real costs or prices used in Figures 6.5 and 6.6 are attempts to measure the "supply price" of coal; hence the use of pithead prices and of an output price deflator. Costs to consumers, such as homes, industries, or electric utilities, require the construction of a "demand price." Such a price, using the consumer price index as a deflator, is intended to reflect the delivered price of coal. Figure 6.7 presents demand prices at the point of consumption since 1700. The price pattern falls somewhere between those of Figures 6.5 and 6.6. The twentieth century sees some upward growth, but nothing as marked as in Figure 6.5. Moreover, the general level of consumer coal prices in the late twentieth century differs little from that for most of the eighteenth

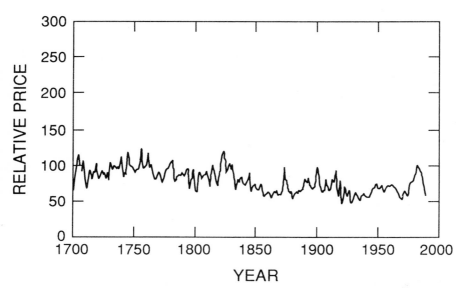

FIGURE 6.6 Relative supply price of coal at the pithead, 1700-1989, deflated by average wage rates.

century. Such a correspondence hardly indicates a worrying trend towards exhaustibility, at least so far as consumers are concerned.

Linking indexes of prices or costs over more than 250 years is a hazardous procedure. It is necessary to weight the data in various ways, which introduces the possibility of error, and the data themselves vary in reliability. Drawing strong conclusions from these data alone is therefore not warranted. At the same time, however, any attempt to gauge long-term exhaustibility of coal ought at least to take the comparisons between the eighteenth and twentieth century data into account.

The static consideration of costs in different collieries at particular times (Figures 6.1, 6.2 and 6.3) shows the cost pattern that would be expected from the operation of diminishing returns. Real costs and real supply prices over long periods of time (Figures 6.4 and 6.5) confirmed the sudden, late rise in prices consistent with the Hotelling model. But the wage-deflated price in Figure 6.6 and the demand price in Figure 6.7 indicate little of it. How can this conundrum be resolved, and why are diminishing returns not always in evidence?

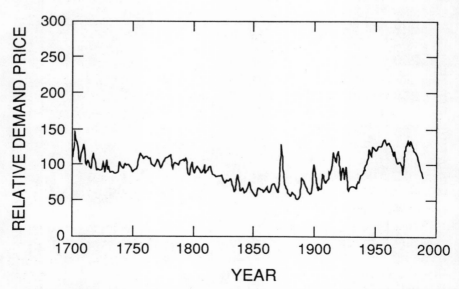

FIGURE 6.7 Relative demand price for coal at point of consumption, 1700-1989, deflated by retail prices (1900=100).

Four arguments are made below that attempt to explain the divergence in supply and demand prices; they treat reducing unit costs, the impact of safety standards, new discoveries, and innovation. None of the four is surprising. Indeed, most of them were conjectured by Jevons, although he rejected each of them in turn. More surprising is the extent to which they clash with the existing theory of exhaustible resources. Therein lies the message of this analysis.

Reducing Unit Costs

The simplest explanation is that costs of marketing and distributing coal fell, hence accounting for the divergence between the supply and the demand price. This trend did occur, but it fails to explain why demand prices failed to rise greatly or, in other words, why the supply price did not rise faster as one moved further into the arena of diminishing returns.

The most direct way for a coal mining firm to keep costs down was to cut the amounts paid per hour for labor. Items such as royalties paid to landowners fell steadily as a percentage of total costs, but rates paid for many factors such as capital were dictated by market trends and were scarcely amenable to straightforward reduction. Wages, which made up the great majority of costs, comprised the largest opportunity to cut costs. Arguments over rates of pay and hours of work became endemic in British coal mining by the early twentieth century. The slogan of "Not a penny off the pay, not a second off the day" took the industry by way of a lockout into the great Coal Strike and General Strike of 1926. Coal mining became notoriously the most strife-ridden industry in Britain in the twentieth century, accounting for a major proportion of all strikes. Strikes in turn became known as the "British disease." Coal mining has always been a comparatively well-paid occupation. But the work has also traditionally been exceptionally onerous and exceptionally dangerous, causing both illness and death. And this is quite aside from whether one wants to spend much of a lifetime several thousand feet underground. It is thus difficult to judge whether the rates of pay offered or requested in any dispute were too high or too low.

What is less contestable is that the opportunity to lower wages, especially in times of collapsing demand, too often provided an easy escape for the colliery owners and managers, allowing them to cut costs without undertaking the radical overhaul of their industry that was really required (Supple 1987). In the late nineteenth and early twentieth centuries, wages in many districts were set by sliding scales that adjusted wages up or down according to price movements. These scales have recently been reassessed by John Treble (1987 and 1990) and found less disadvantageous to workers than often supposed. But one might ask whether these automatic adjustments in wages reduced incentive for radical change. During the period, supply prices began to drift upwards, but also new, potentially cost-saving technology became available both in Britain and overseas, had the owners and managers been sufficiently inclined to adopt it.

Safety Standards

Managers sometimes blamed the imposition of underground working safety standards for the lack of efficiency growth in the industry (Mitchell 1984). The role of government health and safety regulations in any industry remains controversial to our day. Some argue that they preempt technological change and that consumers and producers would best be left to set their own standards (see, for example, Peltzman 1973, for an analysis of the drug industry). Others assert that tough safety and environmental controls generate technological responses from producers and may eventually improve the position of the industry in export markets. Certainly, safety standards for coal mining interacted with technological and organizational change in the industry, though more was achieved through self-regulation, as the industry devised its own solutions, than through externally imposed regulation. There is also little doubt that the track record of the coal mining industry left much to be desired in terms of safety attainments.

Safety standards had an important impact on the extent of exploitation of coal reserves. The increasing depth of worked coal seams in the late eighteenth and early nineteenth centuries led to steadily mounting fatality rates (Hair 1968). The most pressing problem was the incidence of the gases known as "firedamp" and "chokedamp," which contributed, respectively, to explosions in the mines and the suffocation of miners. Accidents reached a crisis point late in the Napoleonic Wars, leading to inventions of safety lamps by Clanny and George Stephenson, and especially by Sir Humphrey Davy in 1815. Although Davy's lamp became celebrated as one of the great inventions of its age, the accepted view of subsequent historians--based on views expressed by the celebrated manager John Buddle to a Select Committee of the House of Commons--was that it did little to reduce fatality rates (Parliamentary Papers 1830: VIIIb, pp. 32-33). Instead it allowed working at still greater depths by offsetting part of the increased risk to miners at those depths--the total risk increased. The important point here is the way in which innovations such as the safety lamp served to offset the effects of diminishing returns but at the cost of such side effects as rising levels of risk.

New Discoveries

A third rather obvious way in which diminishing returns might be offset would be through the discoveries of rich new pits and coal seams. In fact, the regions in which minable coal would be found were by and large known by the mid-nineteenth century. There were a few minor exceptions, such as the long-suspected Kent field in Southeast England, but these turned out to be small. Hence exploration involved less the investigation of wholly new areas than extending the margin of mining at the edges of existing regions. This might involve sinking deeper shafts at existing mines, or moving a little beyond mine boundaries to try to tap suspected seams at another point. With the typical topography of British coalfields being inclined and with heavily faulted seams, these excursions could be a risky and expensive business, so new discoveries cannot explain the lack of diminishing returns.

Innovation

Innovations are the fourth possible explanation for the divergence between supply and demand prices for coal between 1700 and 1989 or the lack of visible diminishing returns in the prices. Conventionally, innovations are divided into process innovations and product innovations. Here, innovations relating to screening coal, preparing coal, coal by-products and distribution are deemed to be product innovations. The process innovations, for the purposes of this industry, are disaggregated into those for access (pit sinking, ventilation, and winding, for example) and those for working (hewing, loading, and underground haulage).

Information about innovations is usually derived from patent data. In England, data on all patents granted are available from the origin of patents in the second decade of the 1600s until 1852 (Woodcroft 1969).

Patents for access technologies in Figure 6.8 show periodic bursts of invention throughout the 234-year period. The peaks correlate reasonably closely with periods of relative shortage of coal and high demand. Innovation was especially needed to solve problems of pumping water out of ever-deepening mines. The solution provided by the steam engine brought a clutch of innovations in the eighteenth century. With the success of the steam engine, the focus of access technologies moved towards the ventilation of mines, as their deepening and widening brought additional problems. These graphs may underestimate the rate of innovation, since many of the significant developments in this sub-field, such as air "coursing" and "splitting," were not patented and probably not patentable.

One may also note a reasonable correlation between access patents in Figure 6.8 and using patents in Figure 6.9. Both fluctuate over time and peak at similar periods, reflecting mainly demand factors. The growing use of coal for steam generation helps give a much sharper increase to patents for using coal than to access patents.

Patents in working (or "getting") coal were unknown in the early part of the period, but assume a rising share of the response to expanding demand in the second quarter of the nineteenth century, as seen in Figure 6.10. Many of the developments of this earlier period, before any mechanization of hewing and loading, were organizational rather than strictly technological--the shift "from bord and pillar" to "longwall" systems of mining, for example. These changes were also not patentable.

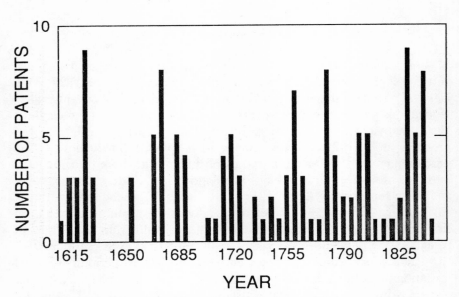

FIGURE 6.8 Patents issued in each five-year interval for access to coal.

After 1852 the British patent data are too ill-organized to permit a ready breakdown, and more impressionistic accounts must suffice. The orthodox story is that the British coal mining industry was plagued by entrepreneurial weakness, exemplified by the slowness to invest in coal-cutting machinery. The argument has been made both for and against this view in a number of recent studies (McCloskey 1971; Greasley 1982 and 1990; Church 1986; and Fine 1990). Although there is some support for the view that geological circumstances (especially thinness of coal seams) accounts for some of the British "backwardness," there are also indications that the industry can stand condemned of entrepreneurial failure in at least four respects.

First was its failure to develop any coordinated approach to improving technology in its major fields of interest. The lack of organized research and development was stressed especially by the Samuel Commission of 1925, but their report failed even at that late date to elicit any response from a fragmented and obsessively competitive industry. Second was the lack of reorganization into larger units, a failure much berated again by the 1925 Commission. The immediate technological effect was that a sizeable number of small mines failed to reach sufficiently large size to justify investment in capital-intensive machinery. In the long run, this hindered

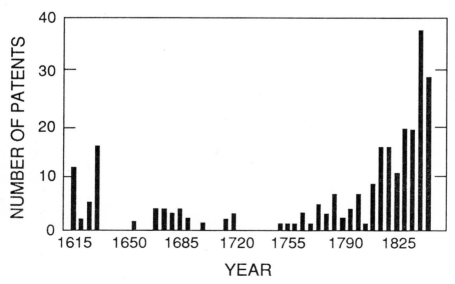

FIGURE 6.9 Patents issued in each five-year interval related to using coal.

productivity growth and competitive strength. A third factor, also associated with lack of reorganization, was the inability to exploit the relationships between innovations in access and in getting coal out of the ground. As the Reid Committee (1945) pointed out, the industry needed not item-by-item replacement of hand hewers by mechanical cutters, but complete updating of the entire system of mining, including locomotive haulage and electrification. The fourth factor was the reliance on keeping down wages to maintain competitiveness, which postponed the need for adopting efficient machinery.

The most convincing reason for the slowness of productivity growth from the second half of the nineteenth century was the general buoyancy of the economic circumstances of mining. Where demand was growing and as yet no significant alternative energy was in view, industry revenues grew. There was consequently little need to improve productivity. Technological advance at best met immediate problems thrown up by diminishing returns, in the manner of crisis management, and costs were passed on to consumers.

By the interwar period the opposite financial conditions prevailed. Demand for coal was now much more elastic, because consumers could use new alternatives such as diesel oil. Also economic disasters depressed

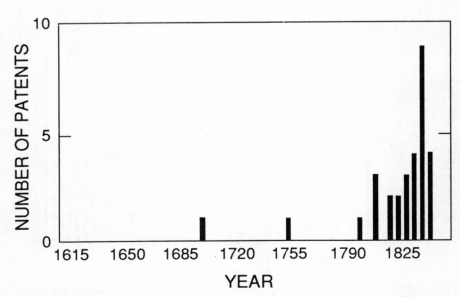

FIGURE 6.10 Patents issued in each five-year interval related to getting coal.

demand. Now the coal industry could claim that it had no resources to permit it to build up technological strength--partly true but also partly a reflection of its earlier inadequacies. Crisis management now prevailed through thick and thin. After World War II, the restructuring of the industry with nationalization produced a brief period of technological success, culminating in the Anderton shearer-loader as perhaps the industry's greatest single advance (Townsend 1976 and 1979; Ashworth 1986). But even nationalization was to become too monolithic and bureaucratic to sustain that success.

If the supply-side responses were insufficient to prevent cost and price increases, as indicated by Figure 6.4, then the demand side had to take up the challenge if real fuel costs were not to rise. Figure 6.11 shows the volume of production of British coal, moving exponentially upwards at virtually constant rates in the eighteenth and nineteenth centuries, and exponentially downwards in unstable fashion in the twentieth.

The patents data in Figure 6.10 show that patents for getting more output from a given quantity of fuel--economizing--rose sharply in the last decade of the eighteenth century. Until then, energy users substituted coal

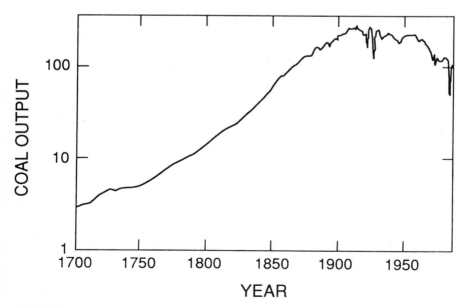

FIGURE 6.11 Growth of coal output, 1700-1989. Output plotted on log scale in millions of tons per year. Data before 1830 are for five-year averages.

and coke for scarcer fuels such as timber. As demand for fuel began to grow very rapidly with industrialization, attention turned from using coal to economizing upon coal.

An early, important innovation that improved coal-use efficiency was the adoption of the Watt steam engine or imitations. The high costs of transportation to London and the manufacturing districts encouraged this substitution, which occurred most rapidly after about 1795, when the license fees for the Watt patent began to decline. In the early nineteenth century, attention shifted to engines of higher pressure, which became the basis of locomotive engines. Another development from the high-pressure engine, one more explicitly aimed at saving fuel, was the compound (two-cylinder expansion) engine. It was adopted in the copper mines of Cornwall because of the relatively high costs of coal in that area, and was further developed in marine engines (von Tunzelmann 1978). After about 1830, the main principles of efficient steam use had been established, but incremental technical change was substantial enough to allow the use of steam power to become truly massive after 1850 (Kanefsky 1979).

More radical responses to reliance on coal consumption came towards the end of the nineteenth century, most notably the steam turbine of Parsons. Use of the turbine dramatically cut manufacturing costs, without necessitating a major reorganization of the factory itself. Slower to develop but ultimately of greater significance was electricity. As Devine (1983) has shown for the United States, the slow adoption of electrification resulted not only from technological difficulties but also from the fact that using electricity efficiently required complete reorganization, changing from a large central engine to individual electric motors on each machine. Finally came the competition from petroleum, although this was largely limited to transportation until after World War II. Coal was never really in contention for automobile fuel; the major substitution of oil for coal in this period occurred in shipping. But the degree of substitution was sufficiently large that, since the late 1950s, coal prices and usage have been dominated by trends in oil.

Jevons (1906) considered all the possibilities for fuel economy. He also considered a remarkably broad range of alternative fuels, including water, tides, geothermal, wood, wind, and even solar energy. But Jevons was not perfectly prescient:

> Petroleum has, of late years, become the matter of a most extensive trade, and has been found admirably adapted for use in marine steam-engine boilers. It is undoubtedly superior to coal for many purposes, and is capable of replacing it. But then, What is Petroleum but the

Essence of Coal, distilled from it by terrestrial or artificial heat? Its natural supply is far more limited and uncertain than that of coal, and an artificial supply can only be had by the distillation of some kind of coal at considerable cost. To extend the use of petroleum, then, is only a new way of pushing the consumption of coal (W. S. Jevons 1906, pp. 184-85).

Experiments to produce oil from coal in fact continued in Britain in a half-hearted manner for about a century after Jevons wrote, until the discovery of North Sea oil. Jevons's point that petroleum was subject to more evident exhaustibility than coal has continued to be correct, but he quite obviously overlooked the feedback cycle between short-run exhaustion, rising costs, and renewed exploration, as discussed above for coal; hence oil supplies turned out to be vastly more significant than he could have presumed.

His major point, however, was not about substitutability, but about fuel economy. If the steam engine or the iron industry reduced its coal inputs per unit of output, then he believed that total fuel consumption would rise. Implicitly he was assuming that the demand for coal was elastic, so that any reduction in the unit cost of fuel would lead to a more than proportionate increase in consumption. However, he failed to develop his economic theory sufficiently to distinguish between shifts in supply and demand for a single commodity and shifts between different commodities (compare Blaug 1962). As a result, his historical evidence confused the two things. The problem of the late nineteenth century was in a way the opposite to what Jevons assumed--in the short run the demand for coal was rather highly inelastic, because of lack of substitutes.

Paradoxically it was this failing in his economics that generated the result he most feared, a negative impact on the growth of British industry. With rising coal costs, the elastic demand would have fallen considerably. But British industry was too dependent on coal in the nineteenth century to achieve such a result. Consequently, industry was forced to suffer years of "dearness and scarcity," such as 1871-1873, and the beginnings of an upward price trend in real terms. The major technological responses (the steam turbine, electricity, and petroleum) were the result of such short-run dependence and turned the upward exponential of output growth, seen in Figure 6.11, into a downward one. Even before this happened, industrial demand turned to some degree from the expensive large coal that dominated markets up to that time, to the cheap small coal, which could with sufficient preparation be used in gas and electricity works, among other things.

It was thus the "positive check," in Malthusian terms, of higher prices that helped turn industrial demand against coal. This took the form of falling demand arising out of fuel economy and greater demand elasticity as substitutes were developed as economic alternatives. These changes did not mean that the British economy in 1900 was faced with an energy crisis of potentially exhaustible supplies of coal, any more than a century earlier it had faced any analogous crisis from lack of water power (von Tunzelmann 1978). Rather, the advance of science and technology external to coal or hydropower had responded to short-run episodes of fuel shortage by initiating an alternative set of technological paradigms (Dosi 1982). As Jevons rightly insisted, events produced a dynamic response of gradually rising charges feeding back to new discoveries and new technologies, long before any physical resource limit was attained.

The price paid by the economy for the savings in coal consumption and for fuel substitution was in the long run, just as Jevons feared, an erosion of the strength of the British economy. Historians have paid surprisingly little attention to this important role of coal in the British economy; analysts of international economic history have addressed the issue more frequently (see Svennilson 1954). Perhaps the ring of determinism is too strong; yet in a recent paper, Gavin Wright (1990) has shown a similar importance of natural resources to American competitive strength in the late nineteenth and early twentieth centuries and found a similar lack of consideration by most historians of this factor.

It is not my argument that historians have been misguided as they search for entrepreneurial or other factors accounting for British industrial decline. On the contrary, the point is that the coal industry neglected the possible "preventive checks" that might have been undertaken. Bolstered by generous profits in the late nineteenth century, it did too little; undermined by inadequate profits in the first half of the twentieth century, it felt unable to do anything. The Hotelling process, although seeming to emerge at first blush from long-run cost analyses, fails to describe what actually happens, at least over the long time horizon considered here. The process instead begins with such static diminishing returns. As those diminishing returns begin to extend over time, they generate searches for new resources and increase the number of lower-cost sources. Or they may force the introduction of new technologies capable of lowering costs in the entire industry and delay the time at which really drastic diminishing returns set in. These both occurred in the British case, but to an inadequate degree. Too often the entrepreneurs of the industry (if they can be so styled) relied on reducing costs only through lowering wages, and in doing so intensified a crisis in industrial relations that is scarcely resolved in our day.

In his second edition, Jevons wrote, "We are now in the full morning of our national prosperity, and are approaching noon. Yet we have hardly begun to pay the moral and social debts to millions of our countrymen which we must pay before the evening" (Jevons 1906, p. 1). Apocalyptic, perhaps, but in the event all too true.

Jevons's son, H. Stanley Jevons (1915), reached the same conclusion. This Jevons, a professor of economics, also wrote a major treatise on coal. Like the father, the son's predictions were rather impressive, including an early forecast of nuclear power (H. Jevons 1915, p. 798). Filial piety required him to support his parent on most issues, but at one point he diverged:

> [I]n his great work on "The Coal Question" my father did not, I think, give sufficient weight to these other factors. . . . The foregoing observations will make it plain how much the Coal Question is in reality a labour question. . . . The price of coal in the future is likely to be affected in greater degree by the attitude of labour, and by the cost of meeting more stringent legislative requirements, than by the mere cost of deeper working and poorer seams (H. Jevons 1915 pp. 758-61).

Like many present-day issues of the environment and the exhaustibility of resources, and as other chapters in this volume show (especially that by Nitze), what appears at first to be economic and technological questions are in the end questions involving values, human beings, and how we want to live.

References

The data employed in the preparation of the figures by the author are derived from a variety of sources with heavy reliance on published statistical abstracts and published and unpublished analyses of these abstracts. Reports of a number of commissions that were established to study the coal industry supply useful additional data. A data appendix, with detailed listing of sources for the figures, is available from the author or from the editor.

Ashworth, W., and M. Pegg. 1986. *The History of the British Coal Industry, Vol. 50, 1946-1982: The Nationalized Industry.* Oxford: Clarendon Press.

Barnett, H. J., and C. Morse. 1963. *Scarcity and Growth: The Economics of Natural Resource Availability*. Baltimore: Johns Hopkins Press.

Blaug, M. 1962. *Economic Theory in Retrospect*. Richard D. Irwin, reprinted by William Heinemann, London.

Church, R. A., A. Hall, and J. Kanefsky. 1986. *The History of the British Coal Industry, Vol. 3, 1830-1913: Victorian Pre-eminence*. Clarendon Press: Oxford.

Dasgupta, P. S., and G. M. Heal. 1979. *Economic Teory and Exhaustible Resources*. Cambridge University Press: Welwyn.

Devine, W. D., Jr. 1983. From shafts to wires: Historical perspectives on electrification. *Journal of Economic History* 43, 347-72.

Dosi, G. 1982. Technological paradigms and technological trajectories. *Research Policy* 11, 147-63.

Fine, B. 1990. *The Coal Question: Political Economy and Industrial Change from the Nineteenth to the Present Day*. London: Routledge.

Flinn, M. W. and D. Stoker. 1984. *The History of the British Coal Industry, Vol 2, 1700-1830: The Industrial Revolution*. Oxford, Clarendon Press.

Greasley, D. 1982. The diffusion of machine cutting in the British coal industry, 1902-1938. *Explorations in Economics History* 19, 246-68.

_____. 1990. Fifty years of coal-mining productivity: The record of the British coal industry before 1939. *Journal of Economic History* 50, 877-902.

Hair, P. E. H. 1968. Mortality from violence in British coal mines, 1800-50. *Economic History Review* 2nd series, 21.

Brooks, D. B. 1974. *Resource Economics: Selected Works of Orris C. Herfindahl*. Baltimore & London: Resources for the Future.

Hotelling, H. 1931. The economics of exhaustible resources. *Journal of Political Economics* 39, 137-75.

Jevons, H. S. 1915. *The British Coal Trade*. London: Kegan Paul.

Jevons, W. S. 1906 [1865]. *The Coal Question: An Inquiry Concerning the Progress of the Nation, and the Probable Exhaustion of our Coal-Mines*. London and New York: Macmillan.

Kanefsky, J. 1979. Motive power in British industry and the accuracy of the 1870 Factory Return. *Economics History Review* 2nd series, 32, 360-75.

McCloskey, D. N. 1971. International differences in productivity? Coal and steel in America and Britain before World War I. In *Essays on a Mature Economy: Britain after 1840*, ed. D. N. McCloskey. London: Methuen. 285-309.

Nordhaus, W. 1973. The allocation of energy resources. *Brookings Papers on Economic Activity*. 529-76.

_____. 1979. *The Efficient Use of Energy Resources*. New Haven: Yale University Press.

Parliamentary Papers. 1830a. VIII. Report of the Select Committee on the Coal Trade.

_____. 1830b. VIII. Report of the Select Committee of the House of Lords, appointed to take into consideration the State of the Coal Trade in the UK.

Peltzman, S. 1973. An evaluation of consumer protection legislation: The 1962 drug amendments. *Journal of Political Economy* 81, 1049-91.

Peirce, W. S. 1989. Rent and technological change in the extractive industries. In B. Carlsson, ed. *Industrial Dynamics: Technological, Organizational, and Structural Changes in Industries and Firms.* Boston: Kluwer, 193-210.

Reid Committee. 1945. Coal Mining: Report of the Technical Advisory Committee, Command Paper 6610, Parliamentary Papers 1944/5, IV.

Samuel Commission. 1926. *Report of the Royal Commission on the Coal Industry, Vol. 1.* London: His Majesty's Stationery Office.

Simon, J. 1981. *The Ultimate Resource.* Princeton: Princeton University Press.

Supple, B. 1987. *The History of the British Coal Industry, Vol. 4, 1913-1946: The Political Economy of Decline.* Oxford: Clarendon Press. Oxford.

Svennilson, I. 1954. *Growth and Stagnation in the European Economy.* Geneva: Economic Commission for Europe.

Townsend, J. 1976. Innovation in coal mining machinery: The Anderton Shearer Loader--the role of the NCB and the supply industry in its development. *Science Policy Research Unit Occasional Paper* ser 3.

_____. 1979. Innovation in coal-mining machinery: The case of the Anderton Shearer Loader. In K. L. R. Pavitt, ed. *Technical Innovation and British Economic Performance.* London: Macmillan.

Treble, J. G. 1987. Sliding scales and conciliation boards: Risk-sharing in the late nineteenth-century British coal industry. *Oxford Economic Papers* 39, 679-98.

_____. 1990. The pit and the pendulum: Arbitration in the British coal industry, 1893-1914. *Economic Journal* 100, 1095-1108.

von Tunzelmann, G. N. 1978. *Steam Power and British Industrialization to 1960.* Oxford: Clarendon Press.

_____. 1979. Trends in real wages, 1750-1850, revisited. *Economics History Review* 2nd ser, 32, 33-49.

_____. 1986a. Malthus's "Total Population System": A dynamic reinterpretation. In D. Coleman and R. Schofield, eds. *The State of Population Theory: Forward from Malthus.* Oxford: Blackwell, 65-95.

_____. 1986b. Steam power and nuclear power: The social savings of British energy technologies. Unpublished.

_____. 1989. The supply side: Technology and history. In B. Carlsson, ed. *Industrial Dynamics: Technological, Organizational, and Structural Changes in Industries and Firms.* Boston: Kluwer, 55-84.

_____. 1990. Cliometrics and technology. *Structural Change and Economic Dynamics* 1, 291-310.

_____. 1991. Malthus's evolutionary model, expectations, and innovation. *Journal of Evolutionary Economics,* in press.

Woodcroft, B. 1969 [1854]. *Alphabetical Index of Patentees of Inventions.* New York: Augustus M. Kelley.

Wright, G. 1990. The origins of American industrial success, 1879-1940. *American Economic Review* 80, 651-68.

PART FOUR

The Environment Goes Global: Issues of the Late Twentieth Century

7

Global Climate Change

John Firor

Local air pollution has been an annoyance and sometimes a health hazard for centuries. And with the emergence of human-produced acid rain and the remote effects of urban smog, it became clear that advanced industrial societies could damage the air over large regions. All of these problems were difficult to manage but seemed susceptible to control measures designed to remove from industrial effluents certain accidental constituents of the waste stream, such as soot, sulfur dioxide, or nitrogen oxides.

In the late twentieth century, however, two atmospheric problems of a different character emerged. They are truly global in extent, more fundamental and less accidental in their cause, and larger in potential impact on society than all the previously studied air pollution episodes. One of these problems--the destruction of stratospheric ozone by synthetic chemicals released worldwide--has been recognized, and international agreements have been signed that may reduce the severity of the problem sometime late in the next century. Many details remain to be understood about this change in the stratosphere, such as the nature and extent of ecological damage that will be produced by the extra ultra-violet light admitted by the thinned ozone layer, but the process of eliminating emissions of the known offenders is proceeding as rapidly as can be expected in the international arena.

This chapter describes the other world-wide atmospheric problem and the processes by which human activities are bringing it about: human-induced global climate change.

Heat Trapping and Global Climate

It has long been understood that the characteristics of the earth's climate are largely determined by certain gases occurring in the air in very small concentrations (see, for example, Dickinson 1982). These gases include water vapor, carbon dioxide, methane, nitrous oxide, and a number of less important substances (Ramanathan et al. 1985). Heat, or infrared radiation, that would otherwise escape from the atmosphere into space is absorbed by these gases. They thereby warm the surface of the earth to an extent that creates here on earth the locale for the evolution and proliferation of a remarkable variety of life-forms. The amount of this warming is large--if all heat-trapping gases could be removed from the air with no other changes, the earth's surface would be $33°C$ colder. The role of these gases has been important in the earth's climate for billions of years, and humans today depend on them to keep the earth habitable. This trapping of heat by atmospheric gases has been recognized and studied for more than a century, and it is one of the better understood features of the atmosphere and climate.

The problem of climate change arises because the concentration of most of these gases is now steadily increasing, raising the prospect of a warming and changing climate. Furthermore, these naturally occurring gases are being joined by some new, synthetic gases that are extremely effective in trapping heat, and that are increasing rapidly in the atmosphere. Climates are known to have changed gradually over geological times, and ecological systems have slowly evolved to adapt to the changing circumstances. But if the changes produced by these rising gas concentrations happen too rapidly for such evolution to take place, ecological and human systems could be disrupted or damaged.

All of these gases differ in their current atmospheric concentration, in their infrared-trapping effectiveness, and in their rate of increase. Thus they also vary in their importance to the normal climate and in their importance to human-induced climate change.

Carbon dioxide is emitted whenever fossil fuel is burned and when forests are cut down and not allowed to regrow. It is the gas producing the strongest influence towards a changing climate, and about six billion tons of carbon in the form of carbon dioxide are released to the air each year as fossil fuel is converted into useable energy. An increased atmospheric concentration that results from this release then continues for a century or more. The annual input to the air from deforestation is more difficult to estimate but probably amounts to one or two billion tons (IPCC 1990).

Methane is released to the air whenever organic material decays without sufficient oxygen, as in rice paddies and the intestines of cows, and it is emitted from animal wastes. This gas is therefore strongly coupled to human agricultural activities. Methane is also emitted by municipal waste dumps and sewage treatment plants; some methane leaks to the air from natural gas wells and pipelines; and coal seams release trapped methane when they are opened for mining (IPCC 1992). The increased atmospheric concentration of methane resulting from human activities is believed to have a potentially shorter duration than that of carbon dioxide--if all extra releases of methane could be suddenly terminated, the atmospheric concentration would likely return to near the pre-industrial level in a few decades, rather than the few centuries needed for carbon dioxide.

Both of these heat-trapping gases--carbon dioxide and methane--have large sources unrelated to human activities and equally large processes that remove them from the air, maintaining an approximate balance over long periods of time. Human activities are now adding new sources that are not completely balanced by increased removal, thereby increasing the amount in the air each year. It is also possible that human activities are changing the processes by which these gases are naturally removed from the atmosphere, thus speeding or delaying the increase in atmospheric concentrations. Methane, for example, is removed in part by reactions with the hydroxyl radical, which is created by photo-induced reactions involving sunlight and water vapor. Carbon monoxide also reacts with the hydroxyl radical, so the emission of carbon monoxide from forest fires and automobiles can deplete the supply of hydroxyl radicals and thus slow the removal of methane. Conversely, extra ultra-violet light admitted to the lower atmosphere by the human-induced destruction of stratospheric ozone can generate additional amounts of the hydroxyl radical and speed methane removal (Madronich and Granier 1992).

Nitrous oxide is released in the agricultural cultivation of soils and in biomass burning. Industrial sources include the manufacture of nylon and nitrogen fertilizer. Like carbon dioxide and the chlorofluorocarbons, increases in the concentration of this gas will remain in the air for a century or more after the termination of excess emissions (IPCC 1992).

The new chemicals being released present a different picture--they were completely absent from the preindustrial atmosphere so they have not been incorporated into some natural material cycle. Halocarbons are chemicals synthesized for use in refrigerators and air conditioners, the blowing of foam insulation, cleaning of industrial products, fire extinguishers, and other uses, and are released to the air when the foam is destroyed, the equipment repaired, or the fire extinguisher discharged. Among these chemicals are the gases that are the chief source of the destruction of stratospheric ozone. Each molecule of these gases is also a very effective absorber of infrared radiation, absorbing up to 30,000 times as much infrared radiation as a molecule of carbon dioxide. Even though the concentration of these gases is much smaller than that of the other gases, they can contribute appreciably to human-induced climate changes, especially changes in the vertical distribution of temperature in the atmosphere. Despite international agreements to phase out the use of the chemicals most responsible for stratospheric ozone destruction, the amount now stored in refrigerators, air conditioners, foam insulation, and fire extinguishers is sufficient to continue sizable releases to the air for many years to come. That, plus the long lifetime of these gases in the air, insures a continuing role for them in human-induced climate change (WMO 1985).

The most important gas in concentration and contribution to the normal climate heating is water vapor. It is both an effective infrared absorber and occurs on average in concentrations of a few thousand parts per million by volume (ppmv) in the atmosphere, the most of any of the infrared-trapping gases. It is not a direct contributor to human-induced climate warming because the amount of water vapor in the air is limited by natural processes. At a given temperature, excess water is rained or snowed out of the atmosphere. Thus human activities have little direct effect on the concentration of water in the atmosphere, except in areas in which an artificial lake or an irrigation project may increase the local relative humidity. But water vapor can have a large indirect effect on climate warming, once temperatures begin to rise. A warmer ocean will evaporate more water into the air, and warmer air can hold more water vapor--increasing the average amount of heat-trapping. Thus water vapor is both an important contributor to the natural climate and will accelerate the climate change driven by human-caused emissions of carbon dioxide, methane, and the other heat-trapping gases.

The ability of humans to modify something as vast as the entire atmosphere arises from two features of human life on earth. The first is that there are so many of us today--5.4 billion in 1992--and we use natural resources so avidly, that our emissions to the air are very large; they are measured (except for halocarbons) in millions or billions of tons each year. The second reason is that the heat-trapping gases are very effective. The sum of all the human-controlled infrared-trapping gases--carbon dioxide, methane, halocarbons, nitrous oxide, and the long list of other, less important, gases is less than one tenth of one percent of the atmospheric volume. Yet these gases are critical in determining the surface temperature of the earth and the distribution of temperature with altitude within the atmosphere (Dickinson and Cicerone 1986).

A new fact has emerged in recent years that both complicates the understanding of the role of each gas in climate heating and illustrates the rapidly expanding ability of people to change the atmosphere. The major atmospheric problems do not act separately but are interactive, one with another. For example, the extra sulfate in the air that causes acid rain, some of it the result of coal burning and metal smelter operation, may induce extra cloudiness, reflect additional sunlight, and thereby slow climate heating, or at least shift the locations of regional climate changes (IPCC 1992). As mentioned earlier, it is also possible that the loss of ozone in the stratosphere can result not only in damage to biological systems at the surface but also in the destruction of methane in the atmosphere, thereby slowing the rate of climate heating.

The atmospheric problems are also linked through common sources. Fossil fuel burning contributes to climate heating, acid rain, and urban smog. Halocarbons destroy stratospheric ozone and participate in infrared-trapping in the lower atmosphere. Agricultural practices encourage methane and also nitrous oxide emissions. The situation is indeed complex and interactive.

Changes in Atmospheric Composition

Carbon dioxide is the most important gas contributing to an enhanced heating of the climate, and measurements of the increasing concentration of carbon dioxide in the air are quite conclusive. Daily measurements made at selected locations during the last thirty years show clearly the increase associated with human emissions. They also show such subtle features as the annual cycle of atmospheric concentration as plants take up some of the atmospheric carbon dioxide during the spring and summer and return it to the air in the fall and winter. The measurements show further that the concentration in the northern hemisphere remains slightly above that of the southern hemisphere, consistent with the fact that most emissions of carbon dioxide from human activities occur in the north, and about a year is required for this excess to mix throughout the global atmosphere.

Analysis of air from long ago, trapped in small bubbles in the Antarctic and Greenland ice caps, shows that the increase in carbon dioxide concentration started about two hundred years ago (Neftel et al.

1985; Friedli et al. 1986). At that time the industrial revolution was getting underway, coal use was increasing, and burgeoning populations were clearing away forests in the temperate latitudes.

The next most important gas which is forcing a changing climate is methane. Again, analysis of air trapped in ice bubbles indicates that the increase in this gas began about two centuries ago (Stauffer, et al. 1985). Each methane molecule is capable of absorbing many times as much heat as a molecule of carbon dioxide, and methane concentration in the air is increasing somewhat faster than those of carbon dioxide. But the initial concentration of methane was sufficiently smaller than that of carbon dioxide that its contribution to global warming is smaller. Figure 7.1 shows the 10,000-year record of atmospheric concentrations of carbon dioxide and methane deduced from the ice core record and from modern measurements. This graph also clearly illustrates the fact that for almost ten thousand years--throughout the history of human civilizations--the sources of atmospheric carbon dioxide and methane were closely balanced by processes that removed these gases from the air. World population is also plotted to suggest the connection between the rapid increase in population growth at the time of the start of the industrial revolution and the sharp increase in the concentrations of these gases.

Climate Model Projections

The amount of heating likely to be produced by a certain increase in the concentration of heat-trapping gases can be estimated from rather simple concepts of the atmosphere, and such an estimate was made almost a hundred years ago (Arrhenius 1896). Modern attempts to refine this estimate use as a tool numerical simulations, or models, of the climate. These models represent the atmosphere with as much detail as possible within the constraints provided by the speed and memory size of large computers. Once constructed and tested, the models are used to calculate how sensitive the atmosphere is to the addition of heat-trapping gases (Dickinson 1986). This is a difficult and expensive task, and only a few scientific groups on earth have the size, budgets, expertise, patience, and computing power to attempt such calculations. The dozen or so models that are in use differ one from the other in mathematical formulation, in the relative emphasis of the various physical processes included in the model, and in the manner in which the atmosphere itself is represented. As a result the models differ in many details, and none of them can be considered a completely satisfactory representation of the climate. But all

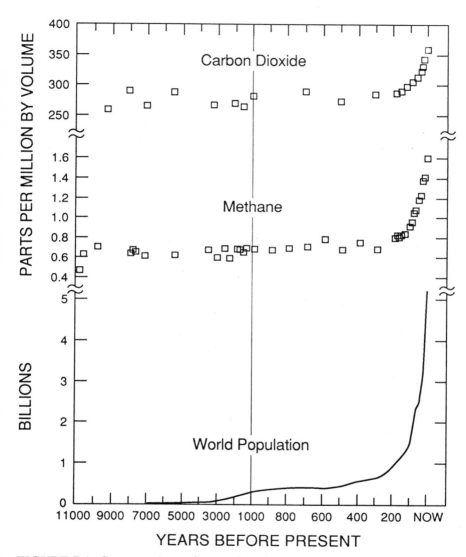

FIGURE 7.1 Concentrations of two infrared-trapping gases in the atmosphere during the past 10,000 years as deduced from measurements made on air trapped in polar ice and from modern air samples. Estimated world population for the same period is shown. Note that the time scale is expanded for the most recent 1,000 years. *Sources:* Carbon dioxide: Keeling et al. 1982; Neftel et al. 1982; Neftel et al. 1985; Methane: Khalil and Rasmussen 1987; Stauffer et al. 1985. Population: McEvedy and Jones 1978; Population Reference Bureau 1992.

of the models agree that the continued increase in infrared-trapping gases in the atmosphere will result in a rapid average heating of the earth's surface and changes in other climate features.

The mean value of the average global surface heating calculated by these models is about 0.3°C per decade, with different models giving values between 0.2°C and 0.5°C per decade (IPCC 1990). There is no reason to suppose that the normal year-to-year and decade-to-decade fluctuations in global temperature will disappear as the climate shifts, so the induced heating will likely be added to the usual wanderings of the climate around some average. This means that the earth's surface temperature is not expected to warm at a uniform rate but will experience periods, decades perhaps, of very rapid warming and periods of slow or no warming.

The Testing of Models

In the normal course of a scientific research project, a prediction resulting from a calculation would be compared with relevant measurements or observations to see if the understanding encapsulated in the calculation is correct and adequate. But because the large natural fluctuations of climate would, so far, obscure any of the predicted rates of human-induced warming, such a comparison for the enhanced heat-trapping effect is still a decade or more away. Even though definitive comparisons are still in the future, public interest in climate change is so high that information on average global temperature is assembled and analyzed promptly in order to catch the earliest possible indication that the human-induced heating has actually been observed. The analyses of surface temperature measurements for the last century or more from locations around the globe have all shown an average heating of about the expected magnitude (IPCC 1992; Jones et al. 1986). The Intergovernmental Panel on Climate Change summarized this situation in its first report, and repeated it in its more recent summary, as follows:

> . . . the size of this warming is broadly consistent with predictions of climate models, but it is also of the same magnitude as natural climate variability. Thus the observed increase could be largely due to this natural variability; alternatively this variability and other human factors could have offset a still larger human-induced greenhouse warming (IPCC 1990, 1992).

This situation would be frustrating for the scientists but otherwise harmless if the earth's climate were not fundamental to many aspects of society. The social importance of knowing how rapidly the climate will change makes it urgent to have convincing validation or refutation of the model projections of rapid climate change resulting from the added infrared-trapping gases. Scientists studying climate change are therefore under pressure to find ways of determining the reliability of their calculations more prompt than simply waiting for several decades.

Climate models are very elaborate, but even with this complexity it is possible that modelers have failed to consider properly some climate feature of major importance. The models already include secondary processes, or feedbacks, that affect the results strongly, so it is reasonable to ask how certain it is that there are no others that should be included. A large positive feedback mentioned earlier arises because higher temperature will lead to increased evaporation from the oceans and to higher water vapor content of the air. Since water vapor is an infrared-absorbing gas, this additional atmospheric moisture will add to the infrared trapping, further heating the surface. A frequently discussed negative feedback supposes that the additional water vapor will also lead to additional cloud cover, which in turn would reflect more sunlight to space, lowering the heat input to the surface, and thereby decreasing the net amount of heating. Both of these processes are included in climate models, but the cloud feedback process is only crudely simulated. Since many of the cloud measurements needed to improve the cloud simulations are lacking and will take years to achieve, and since it may never be possible to assert that all possible feedbacks have been considered, techniques other than separately confirming that each portion of the model satisfactorily represents some process in the atmosphere must be used to build confidence in the current calculations.

The verification techniques actually employed involve comparisons of climate model calculations with observations of present and past climates and climate changes. The models can be used, for example, to calculate the observed climate of today and the results compared in detail with actual observations. The models further can be used to simulate the difference between average July weather and average January weather--a brief but very large climate change--and again compared with the observations. Certain large scale correlations, such as the large-scale weather patterns that develop during an El Niño event--a period during which warmer-than-normal sea water appears off the equatorial Pacific coast of South America--can also be modeled and compared with what actually happens. None of these comparisons tests the model in exactly the mode in which it is employed while projecting climate heating, but each such test successfully passed reduces the chance that some important process has been omitted from the model. And, indeed, the large, three-dimensional, global climate models compare reasonably well with actual measurements when tested in this manner (Firor 1990; IPCC 1990, 1992).

Today's models cannot be used to estimate the pattern of heating beyond a few degrees, since at some point a major readjustment of the climate system will take place. An ocean current will reverse, for example, or the loss of summer ice in the Arctic Ocean will shift winds throughout the Northern Hemisphere. These major shifts present additional difficulties to the model. As far as policy considerations go, this lack of ability to foresee major readjustments makes matters worse, since a shift of this sort would cause sudden changes that would make the task of adapting to the changing climate even more difficult.

The projected heating rate--$0.3°C$ per decade--contains an assumption about the behavior of people. It was necessary for the derivation of this number to estimate how much the concentrations of infrared-trapping gases in the air will increase. The $0.3°C$ per decade value for the rate of warming is based on the assumption that the world will take a business-as-usual approach to fossil fuel use and to the release of other heat-absorbing gases, except for the chlorofluorocarbons, which are the subject of an international protocol requiring a gradual cessation of their production.

This heating estimate is also based on the assumption that the mechanisms that govern the emission of infrared-trapping gases by natural systems and the mechanisms that control the cleansing of such gases from the atmosphere will not change as the global temperature increases. This assumption is the subject of considerable controversy. As the surface waters of the ocean warm they will be able to dissolve less carbon dioxide, so the ocean uptake of a portion of the human emission could be reduced. The emission of carbon dioxide by forests--directly from trees as a result of their metabolic processes, and from the decay of organic material in the forest soils--is known to increase at higher temperatures. Each of these processes could increase the emissions of carbon dioxide. Conversely, the effect of extra carbon dioxide in the air around individual plants has been shown to increase plant growth, so it is possible that plants everywhere are taking up some of the extra carbon dioxide and will continue to do so. It is also speculated that a warmer ocean or melted permafrost on land could emit extra methane now held on the bottom of shallow, cold seas or incorporated into the frozen soil.

In addition to these uncertainties about future atmospheric concentrations of infrared-trapping gases, it would raise confidence in the calculations if the individual models agreed more closely on the predicted rate of heating than the current range of 0.2°C to 0.5°C per decade. Furthermore, though calculations of changes in the global average temperature are convenient for modelers and provide an overall index for monitoring observations, global average temperature is of no help to anyone planning measures to adapt to the change. Estimates of future conditions--temperature, precipitation, winds, cloudiness--at each particular location will be needed for such planning. Currently the models lack such detail, and reducing this limitation will require improved descriptions of some of the physical processes in the models and the availability of larger computers.

Impacts of Climate Change

Most models agree on the continental-scale distribution of the climate heating. They foresee that land will warm more rapidly than the ocean; in winter, high northern latitudes will warm more than the global mean; and temperature increases in southern Europe and central North America will be larger than the global average, accompanied by reduced summer precipitation and soil moisture (IPCC 1990).

Increased evaporation that will also accompany higher temperatures could decrease the average flow in some river basins (see Gleick, this volume). One effect that will be truly global, and hence not suffer from the lack of spatial detail in the models, will be the rise in sea level--eventually averaging five centimeters per decade--that will push estuaries inland in competition with other land uses and contribute to damage as each storm arrives at the coastline on a "higher tide" than previously. There will be apparent regional variations in the rate of sea level rise due to rising or subsiding motions of islands and continental margins or to shifts in the mean surface winds in coastal regions.

One method of estimating the total societal impact of an average warming of one-third degree per decade is to examine climate changes in the past and see what impact they had. During the last 1,000 years, as shown in Figure 7.2, the earth has been warmer than now, around the 11th Century, and colder, during the "Little Ice Age" a few centuries ago. It is estimated that each of these periods represented a widespread--perhaps global--change in average temperature of about one half degree C. These changes appeared and disappeared at rates of perhaps .05° C per decade.

These apparently small, slow changes forced shifts in agricultural practices and resulted in the destruction of marginal societies (Rotberg and Rabb 1981). Based on this observation one should expect that a climate change occurring more than five times as fast could have major impacts on today's society, especially if the change is allowed to continue until late in the next century when it could be five times as large as these earlier events, and still increasing. In Figure 7.2 the dashed lines extending into the future show, to the same scale, warmings of 0.2°C and 0.5°C per decade--the range of possible warmings deduced from a survey of all the model calculations. As mentioned earlier, it is not expected that this warming will be

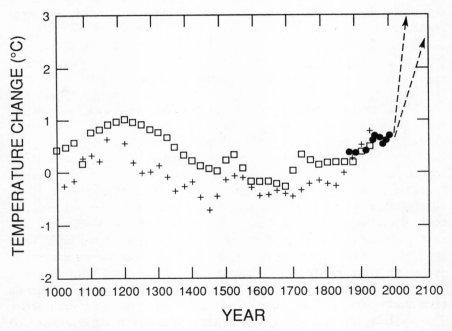

FIGURE 7.2 Surface temperatures during the last 1,000 years. Squares represent data from central England (Lamb 1977); crosses show tree ring widths from trees in the White Mountains in California believed to be sensitive to summer temperature (La Marche 1978); and the solid circles represent modern measurements of the average temperature of the northern hemisphere (Jones et al. 1986). The scale indicating degrees C of temperature change is based on the central England measurements and the modern observations. The vertical scale for the tree ring data has been adjusted to show fluctuations of a similar amplitude. The dashed lines at the right end indicate the range of warming projected by climate models (0.2 to 0.5 °C per decade) for a "business-as-usual" scenario of heat-trapping gas emissions (IPCC 1990).

steady, as implied by the straight dashed lines, but will have periods of more rapid temperature increase and other periods of slower warming.

Such a large, rapid change will clearly force major shifts in agricultural practices. Agricultural experts emphasize that agriculture in industrial countries is flexible and innovative, and that many steps can be envisioned that will prepare society for this change--new crop varieties developed, more and better irrigation systems built, improved fertilizer and pesticide formulations created. What is frequently overlooked, however, is that the change will not be a shift from today's climate to a new, hotter, but stable, climate. Rather society must be preparing for a continuing increase in surface temperature to be added to the large year-to-year weather variations that now occur. The adaptation problem therefore is not one of planning for a new climate but instead planning for a new climate every few decades. Thus adaptation means developing a continuous stream of new crop varieties; building new irrigation systems where dryer climates will prevail, and bringing water, in increasing quantities, from wherever it is available; and advancing pest management strategies steadily to deal with those warm-climate insects and diseases that will extend their range.

Slowing the Change

Continuing to modify the composition of the earth's atmosphere for the indefinite future is a risky course of action. Yet it is unlikely that society could even contemplate, much less implement, a prompt cessation of the emission of every offending gas. In order to approach a stable atmospheric composition, however, a complete cessation of emissions is not required.

Even though the processes that remove the heat-trapping gases from the air are not today balancing the extra emissions, the higher atmospheric concentrations already occurring have generated some additional cleansing activity. For example, the amount of carbon annually emitted by human activity in the form of carbon dioxide is estimated to be seven or eight billion tons. But the atmospheric content only increases by about 3.5 billion tons yearly (IPCC 1990). Clearly some processes have been activated or accelerated by the higher concentration of carbon dioxide, and these processes are removing and storing annually an amount of carbon that is an appreciable fraction of the carbon that is emitted each year. It seems likely that these extra removal processes will continue for decades or centuries if the atmospheric concentration of CO_2 remains at its present high level. Therefore reducing emissions to a value equal to the extra removal rate could produce an approximately constant concentration of carbon dioxide in the atmosphere. Similar calculations can be made for the other heat-trapping gases.

The current estimates of how large a reduction of emissions would be required to stabilize the atmospheric concentration of each gas are as follows (IPCC 1992):

Carbon dioxide: immediate reduction of 60 to 80 percent in order to stabilize at 353 ppmv, or a steady reduction to 50 percent of today's amount by 2050 in order to stabilize at 420 ppmv.

Methane: immediate reduction of 15 to 20 percent. The relatively small reduction required to stabilize this gas results from its short lifetime in the atmosphere.

Halocarbons: reductions of 70 to 95 percent, depending on which halocarbon, in order to stabilize and then reduce the atmospheric concentrations below the current levels, which are already observed to be destructive of stratospheric ozone.

Nitrous oxide: reductions of 70 to 80 percent.

Thus stabilizing the atmosphere's concentration of heat-trapping gases requires reduction in emissions ranging from relatively modest values for methane to large values for the other gases. Though far from a trivial task, reductions in these amounts are less onerous than reductions to zero.

Possible Societal Responses

Today, there is a lively scientific effort in place to improve understanding of the processes important to global climate change and to incorporate this understanding into more elaborate, fine-scale climate models. It is expected that this research program will, over the next several decades, improve the ability to foresee the impact of various heat-trapping gases in the atmosphere. In the meantime, people want to know whether human activities are about to produce a climate change of sufficient magnitude and rapidity to require action today in addition to enhanced research.

The judgment of many of those conducting global climate modeling studies, and the judgment of those reviewing and summarizing these studies for governments, is that the likelihood of a rapid warming of the earth's surface and lower atmosphere is sufficiently high to warrant action now, both to reduce emissions of the heat-trapping gases and to improve the capability of people and natural systems to deal with current climate variability (See, for example, NAS 1991). Nevertheless, lively controversy over whether such actions are appropriate continues. It is said that more certainty is required to warrant a response to projected climate change and that more precision is required to fashion the most appropriate actions.

But from the point of view of leaders striving to solve the problems that are before them, more certainty and more precision about climate change may not be as critical to a decision as they at first appear. For example, carbon dioxide from fossil fuel burning is the main contributor to the projected climate heating. Fossil fuels are also the main cause of urban air pollution, with its toll on human health; they are the cause of human-induced acid precipitation, with its accompanying damage to buildings, paint, bridges, trees, and lakes; they contribute to low altitude ozone and its damage to agriculture, forests, and children's lungs; they impair visibility over national parks and other scenic vistas and over airports; their import contributes heavily to U.S. balance-of-payments and debt problems; and their mining, drilling, and transport contribute oil spills and degraded landscapes to the list of national problems.

In this context, it is not clear that one needs to know whether the climate will heat at $0.2°C$ per decade or $0.5°C$ per decade before deciding to reduce reliance on fossil fuels. More broadly, it is thought that the cumulative impact on the environment of a rapid climate change may in a few decades produce disturbing changes in ecosystems. But already, world wide, forests are disappearing at an acre a second and in the process indigenous people are being displaced and their culture destroyed (Nations, this volume). Thus a decision to lessen total human impact on the natural world depends on more than projections of climate change. From this point of view, climate change is just one more indication of the need to effect major changes in how we manage our affairs. Some of these indications are certain; some are only partly understood. But the total is impressive (See, for example, Speth, this volume).

It is also suggested that we need more certainty about climate change in order to design some sort of technological response. If we knew exactly how fast the climate will get hotter, for example, perhaps we could calculate how many billions of small, silver balloons to float around the earth in the stratosphere to reflect the right amount of sunshine to get rid of the right amount of heat to keep the average temperature the same (NAS 1991).

There are two difficulties with this engineering approach to countering human-induced climate heating. First, society already has, on the shelf, technological means of slowing the heating, means requiring much less organizational skill than a global balloon system. These technologies, which improve the efficiency of energy generation and use, allow all of today's activities to be conducted while using much less fossil fuel (Goldemberg et al. 1987a, 1987b; Chandler et al. 1988; NAS 1991).

The second difficulty with an engineering approach is that it is not likely to succeed in eliminating important local human-induced climate changes. The balloons will reflect heat on average at a different place on earth from where the carbon dioxide traps heat, so the net result could be to exchange a climate change characterized by an average global heating for one with a heightened set of rapidly changing regional climate anomalies. This objection is not limited to the balloon proposal. If society sets out to cancel a strong force on the climate, such as the one created by adding extra heat-trapping gases to the atmosphere, it would have to create another equally strong force, one that would have its own set of impacts on ecosystems and on society.

Conclusions

Current calculations indicate that a continuing addition of infrared-trapping gases to the atmosphere will result in a rapid heating of the earth's surface and in changes in other features of regional climates. These climate shifts are likely, in concert with other human environmental impacts, to damage natural ecosystems and disrupt social and political systems. Though the science of a projected climate heating is not perfect, it is good enough to warrant concern. A possible reaction to this concern is to reduce heat-trapping gas emissions, and the means exist to reduce emissions of many of these gases through efficiency improvements. Emissions reduction can slow climate change without introducing other, potentially troublesome, changes in natural systems, and perhaps reduce the total impact of human populations and their activities on the earth.

References

Arrhenius, S. 1896. On the influence of carbonic acid in the air upon the temperature of the ground. *The London, Edinburgh, and Dublin Philosophical Magazine and Journal of Science* April 1896, 273-285.

Chandler, W. U., H. S. Geller, and M. R. Ledbetter. 1988. *Energy Efficiency: A New Agenda.* Washington: American Council for an Energy-Efficient Economy.

Dickinson, R. E. 1982. Modeling climate changes due to carbon dioxide increases. In *Carbon Dioxide Review 1982,* ed. W. C. Clark. New York: Oxford University Press.

———. 1986. How will climate change? In *The Greenhouse Effect, Climate Change, and Ecosystems,* ed. B. Bolin, et al. SCOPE 29. New York: John Wiley and Sons.

Dickinson, R. E., and R. J. Cicerone. 1986. Future global warming from atmospheric trace gases. *Nature* 319, 109-115.

Firor, J. 1990. *The Changing Atmosphere.* New Haven: Yale University Press.

Friedli, H., H. Lotscher, H. Oeschger, U. Siegenthaler, and B. Stauffer. 1986. Ice core record of the $^{13}C/^{12}C$ ratio of atmospheric CO_2 in the past two centuries. *Nature* 324, 237-238.

Goldemberg, Jose, T. B. Johansson, A. K. N. Reddy, and R. H. Williams. 1987a. *Energy for a Sustainable World.* Washington: World Resources Institute.

———. 1987b. Energy for Development. Washington: World Resources Institute.

IPCC [Intergovernmental Panel on Climate Change]. 1990. *Climate Change: The IPCC Scientific Assessment,* ed. J. T. Houghton, G. J. Jenkins, and J. J. Ephraums. Cambridge: Cambridge University Press.

———. 1992. *Climate Change 1992,* ed. J. T. Houghton, B A. Callander, and S. K. Varney. Cambridge: Cambridge University Press.

Jones, P. D., T. M. L. Wigley, and P. B. Wright. 1986. Global temperature variations between 1861 and 1984. *Nature* 322, 430-434.

Keeling, C. D., R. B. Bacastow, and T. P. Whorf. 1982. Measurements of the concentration of carbon dioxide at Mauna Loa Observatory, Hawaii. In *Carbon Dioxide Review, 1982,* ed. W. C. Clark. New York: Oxford University Press.

Khalil, M. A. K., and R. A. Rasmussen. 1987. Atmospheric methane: Trends over the last 10,000 years. *Atmospheric Environment,* 21, 2445-52.

Lamb, H. H. 1977. *Climate: Present, Past and Future,* v 2. Climatic History and the Future. London: Methuen.

La Marche, V. C. 1978. Tree-ring evidence of past climatic variability. *Nature* 276, 334-38.

Madronich, S., and C. Granier. 1992. Impact of recent total ozone changes on tropospheric ozone photodissociation, hydroxyl radicals, and methane trends. *Geophysical Research Letters* 19, 465-67.

McEvedy, C., and R. Jones. 1978. *Atlas of World Population.* New York: Penguin Books.

NAS [National Academy of Sciences]. 1991. *Policy Implications of Greenhouse Warming.* Washington: National Academy Press.

Neftel, A., H. Oeschger, J. Schwander, B. Stauffer, and R. Zumbrunn. 1982. Ice core sample measurements give atmospheric CO_2 content during the past 40,000 yr. *Nature.* 295, 220-23.

Neftel, A., E. Moor, H. Oeschger, and B. Stauffer. 1985. Evidence from polar ice cores for the increase in atmospheric CO_2 in the past two centuries. *Nature* 315, 45.

Population Reference Bureau. 1992. *1992 World Population Data Sheet.* Washington: Population Reference Bureau.

Ramanathan, V., R. J. Cicerone, H. B. Singh, and J. T. Kiehl. 1985. Trace gas trends and their potential role in climate change. *Journal of Geophysical Research* 90, 5547-66.

Rotberg, R. I. and T. K. Rabb, eds. 1981. *Climate and History.* Princeton: Princeton University Press.

Stauffer, B., G. Fischer, A. Neftel, and H. Oeschger. 1985. Increases in atmospheric methane recorded in Antarctic ice core. *Science* 229, 1386-1388.

World Meteorological Organization. 1985. Global Ozone Research and Monitoring Project, Report 16, Vol. 1, pp 57-77. Geneva.

8

Global Water Resources: The Coming Crises

Peter H. Gleick

The Greek poet Pindar said, "Water is the best of all things." This was at a time when the known elements were earth, wind, fire, and water. We know of more elements now, but water is still wonderful stuff. A little bit of hydrogen, a touch of oxygen, and you have fresh water--the difference between a planet humans can live on and one they cannot.

Already a critical element of life on Earth, fresh water is likely to become more critical in the next few decades. Issues of fresh water quality, fresh water quantity, and conflicts over water resources shared between countries may well overshadow many other environmental problems in coming years, including some now considered critical. These water problems are symptoms of a larger set of other problems: growing population, global climate change, resource disparities between rich and poor nations, and widespread resource mismanagement. Thus, in addition to traditional water pollution problems, such as those discussed at the first United Nations Conference on the Environment in Stockholm in 1972, it is necessary two decades later to address a new set of important questions. How will different paths to sustainable economic development affect water supply, demand, and quality? How will global climate change affect water resources? How will conflict over shared international water resources arise and how might we develop mechanisms to prevent and resolve those sorts of conflicts? After a brief discussion of the physical characteristics of global water resources, each of these issues is considered in turn.

Ninety-seven percent of all water on Earth is salt water. Though many problems afflict the seas, the focus of this chapter is the other three percent of global water resources--fresh water. This three percent amounts to about 30 million cubic kilometers, but a vast portion of that volume is locked up in the ice caps or in deep ground water reservoirs. Most of this fresh water is therefore effectively beyond human reach, economically or physically. About 200,000 cubic kilometers of water is readily available in lakes, rivers, and shallow ground water, or is raining out of the sky (see Shiklomanov 1992).

That 200,000 cubic kilometers of water is not evenly distributed, either in time or space. People do not necessarily get water when or where they want it. We therefore spend hundreds of millions of dollars worldwide annually to build reservoirs, aqueducts, and other structures to store it, ship it from one part of the world to another, process it, desalinate it, and otherwise overcome the shortcomings of natural water distribution.

For example, runoff per capita in Africa is two-thirds of the U.S. average, while per capita runoff in South America is three and a half times the U.S. average (World Resources Institute 1990). Disparities in water availability also exist at smaller scales. In the United States, fresh water runoff on average amounts to about 14,000 cubic meters per person per year, or about 10,000 gallons per person per day. But a family of three living in northern California learns from their water bill that they use about 200 gallons per day. Where is their other 29,800 gallons per day? California as a whole does not receive the U.S. average of 10,000 gallons per person per day. Much of California is a desert, and on average it is a lot drier than much of the rest of the country. Average runoff there is only about 2,000 gallons per person per day. Still, where is that family's other 5,800 gallons? Much of it goes to grow crops, to produce electricity, to provide for industrial or commercial needs, or is lost from reservoirs through evaporation. If any water is left, it goes to support natural ecosystems and fisheries; ecosystem needs are typically at the bottom of the list, both in the United States and elsewhere.

Overcoming or coping with uneven water availability is an on-going activity with which the world has a lot of experience. We have less experience with the trio of issues emerging in the coming decades: the role of water in sustainable development, the impact of global climate change on water resources, and conflict over shared water resources.

Water and Sustainable Development

Experts have paid much attention lately to the issue of sustainable development, the notion that a decent standard of living for today's population should be achieved without compromising that standard of living for future generations (see for example WCED 1988). "Standard of living" includes such things as human health, education, work, and other elements of well-being, as well as materials and resource use. The notion of sustainable development must make some assumptions about all of these factors. And water plays a role in each of them.

It is important in discussions of sustainable development to distinguish between renewable and non-renewable resources. To preserve the ability of future generations to have decent lives, non-renewable resources need to be used slowly while economical and non-polluting substitutes are developed. And renewable resources should not be used more rapidly than they renew themselves. In general, water is a renewable resource: using it does not diminish its overall long-term availability and our ability to reuse it. But there are examples of water being used faster than the renewal rate, such as when water is withdrawn from a ground water aquifer faster than it can be naturally recharged, or when we contaminate a water supply, making it unfit for further use.

Ironically, we are now experiencing limits on some renewable resources, like water, faster than we are running into limits on some non-renewable resources. In large part this is so because renewable resources are often considered free goods, and we put few limits on our use of them. In contrast, non-renewable resources are priced and become increasingly more expensive as we consume them (see, for example, von Tunzelmann, this volume). While water is in general renewable, a limited amount of water is available in a region or during a particular period of time. Limits on renewable resources may become a seriously disconcerting problem as the quest for sustainable development is undertaken.

There is a tremendous range of water use worldwide. More than 20 countries use between 1,000 and 5,000 cubic meters per person per year, while over 50 countries use less than 100 cubic meters per person per year (Gleick 1992a). These differences reflect differences in levels of economic development and in actual resource availability. Africa is a water-poor continent: almost all countries of Africa use less than 100 cubic meters per person per year. On the other hand, poor countries may use a great deal of water. Afghanistan, for example, uses well over 1,000 cubic meters per person per year, though it is among the poorest countries in the world.

Water for drinking and sanitation, water for irrigation and agricultural production, and water for energy production are important to economic development. Distressingly, over one billion people worldwide do not have access to what the U.N. considers safe drinking water. The one billion figure includes nearly 50 percent of Africa and 30 percent of Asia (excluding China, for which reliable data are not available) (Christmas and de Rooy 1991).

In the 1980s, the United Nations made a decade-long commitment to provide safe drinking water and sanitation for the world's poor. Major efforts and immense progress were made, but sadly, all that effort and progress barely kept even with population growth. In 1980 over one and a half billion people lacked both safe water and sanitation; in 1990, over a billion still lacked safe drinking water and one and a half billion still lacked safe sanitation, despite the fact that hundreds of millions of people were provided with such access during the decade. Development efforts are in a race against a growing population and are barely keeping up.

Water is also a fundamental factor in providing food for expanding populations. The green revolution of the 1960s-1970s was in many ways a revolution in water supply for agriculture--the critical elements were new crop varieties, fertilizer, pesticides, and irrigation water. Perhaps 15 percent of cropland worldwide is irrigated, but over 30 percent is irrigated in Asia where the green revolution had the greatest effect (FAO 1990). Irrigated lands also provide much more than 10 percent of total global crop production, and future growth in yields requires either an increase in the total amount of irrigated land or great increases in the efficiency with which irrigation water is used.

Yet providing more irrigation water conflicts with demands for industrial development, commercial water use, and household water use by a growing population. At the same time, continued irrigation often damages cropland, undermining the very increase in food production that it was intended to provide. Water logging and salinization are the principal threats to long-term irrigation. In India, for example, over 40 percent of the irrigated land has been damaged by salinization, meaning that yields from these lands have fallen noticeably (Postel 1989). Salinization is a function of the quality of the water used for irrigation, the quality of the land, and how the water is applied; not all irrigated lands become saline. Still, in many countries--particularly in areas of significant irrigation such as India and Pakistan--increasing damage to irrigated lands will greatly complicate ongoing efforts to improve crop yields.

A final element of sustainable economic development is energy; the world's developing countries need enough energy produced in an environmentally acceptable manner to provide for their needs. As energy is fundamental to development, so water is fundamental to the production and use of energy. There is a great disparity between energy use in the industrialized nations and in the poorer nations. On average, industrialized nations use seven times the energy per person used by developing countries, and they tend to use forms of energy that are water-intensive. The conventional forms of energy production--fossil fuel and nuclear--use far more water per unit of energy produced than alternative renewable and non-conventional energy sources (Gleick 1992b). Water is almost certain to be an important factor in decisions regarding the types of energy used in the future and the types of energy necessary for sustainable economic development.

Global Climate Change

Global climate change will not affect the generally renewable nature of water, but it will dramatically alter water availability, timing, and perhaps most importantly, its predictability. Global climate change will add a great degree of uncertainty to issues of water availability and water resources planning.

John Firor (this volume) explains the fundamentals of global climate change. It is known that greenhouse gas concentrations are increasing and that increasing concentrations of these gases will trap more heat and more energy in the atmosphere. This in turn will lead to changes in temperature and other climatic parameters, though temperature is mentioned most commonly. Among the other climate parameters are several of interest to water resource planners and hydrologists: changes in precipitation patterns, sea level, and storm frequency and intensity.

All of these climate elements will affect water resources as they change: higher temperatures mean greater demand for water; higher temperatures mean greater evaporative losses from land and water surfaces; changes in precipitation may mean receiving water at different times of the year than we are accustomed to; and changes in storm frequency may mean changes in the reliability of monsoons and other large-scale events determining precipitation patterns.

The basic science of a human-induced climate change on a global scale is well understood. A number of uncertainties remain, however, including the nature of climate changes on a regional scale and their implications. Unfortunately, most water resource management takes place on the regional scale, so that current projections of global climate change are only of general interest to water managers.

Some of the difficulties of water planning, given these uncertainties, are illustrated by the 1976-1977 drought in California, until recently the most severe on record. At that time, 1976 was the fourth driest year recorded, and 1977 was the driest year. Water managers had the job of responding to the situation. California water managers repeatedly face a dilemma: the state gets nearly all of its precipitation in the winter season of October to April, and almost none from May to September. As a result, reservoirs are kept low in the winter for flood control. Managers must decide when the risk of flooding is low enough to begin to fill the reservoirs in anticipation of the dry irrigation season. It is best for the reservoirs to be as full as possible entering the dry part of the year.

Exactly when managers decide to start filling reservoirs depends on when they expect to get water, which in turn depends on past experience. The state water managers look at the past record and decide what to do. The risk is that if they make a mistake and fill the reservoirs too soon, parts of the state could have severe flooding; if they start to fill them too late, there may not be enough water to carry the farmers through the dry period. Either of these alternatives is bad for California.

In 1978, following the drought, water managers decided to change their operating rules and fill the reservoirs a bit earlier to ensure that a repeat of the 1976-1977 drought could be handled better. Then in 1983, California had the wettest year on record. With the reservoirs fuller, serious flooding caused hundreds of millions of dollars of damage. The water managers responded by changing the rules back again, filling the reservoirs late in anticipation of high runoffs.

Now California finds itself in the sixth year of an unprecedented drought. Significant amounts of rain fell in March 1991 and in February 1992, but not enough to break the drought state-wide. The 85 to 100 years of reliable rainfall records that exist for California are not enough to tell water managers what to expect next; they do not contain patterns like that of the last six years. The records are also too short to say whether the current drought is the result of a gradual climate change or just normal variation. In either case, not enough is known to tell managers what they say they want: how much water they will receive next year.

The Climate and Water Panel of the American Association for the Advancement of Science recently attempted to address the concerns of water managers about climate change and uncertainty (Waggoner 1990). They went beyond the preparation of a report and set up a tour to talk with local, state, and federal managers about these issues. At those meetings, every water manager said in effect: "Tell us what is going to happen, and we will know what to do about it. Tell us how much water we are going to get, and we can manage it." Unfortunately, while changes in climate patterns are likely, the details of these changes remain obscure.

The growing possibility of a warming climate thus adds another layer of uncertainty to the already uncertain business of managing water. It is no longer possible to assume that the future will look like the past, and it is also not possible to predict that new future precisely. For California, the best course may be to assume greater uncertainties and either accept a greater risk of flooding or drought or change the management of the non-physical parts of the system, such as the operation of water markets and water pricing. The problem of climate change is not something water managers and planners are comfortable with, yet they need to begin seriously considering how to handle it in order to avoid unpleasant surprises in the future.

Conflict over International Water Resources

Many fresh water resources are shared by two or more countries, though Americans are largely unaware of this fact. More than 50 percent of all the land area in the world is drained by rivers shared by two or more nations.

The history of conflict over shared water resources is long, and examples range from the Nile in Africa to the Jordan in the Middle East to the Ganges, Brahmaputra, and Indus in southern Asia. Conflict over these shared water resources is likely to increase in the future for at least two reasons. First, growing populations mean a growing demand for water for all uses but perhaps most importantly for additional irrigation water to increase food production. Secondly, conflict over shared water resources may increase in some regions because of global climate changes that alter either the quality or quantity of water available to users. As conflicts increase, they may also intensify and become more violent (Gleick 1992a).

Population growth is the driving force for greater demand for water, which is needed to supply human needs for drinking water, sanitation, commercial and industrial uses, and irrigation. Places like India, Bangladesh, and Pakistan--with rapidly growing populations--are likely to see rising tensions over limited runoff in international rivers.

Irrigation demands will lead to more and more withdrawals, and more consumptive use of water. The situation on the Nile River is a good example of the limits faced by rising demands. The Nile flows through nine nations in northeastern and north Africa. Egypt is the downstream nation, at the end of the pipe, yet it is the biggest user of Nile River water. Furthermore, it has virtually no alternative sources of supply. Ninety-seven percent of the flow in the Nile is consumed in Egypt, and almost all of this water originates outside of Egypt's borders.

The Sudan, the nation immediately upstream of Egypt, is often considered the nation in Africa with the greatest potential for increasing agricultural production. It has irrigable land, and the Nile flows through its territory. But a treaty signed by Egypt and the Sudan (and none of the other seven nations bordering the Nile) limits the amount of water the Sudan can use, preventing significant further irrigation development. If the Sudan were to try to increase agricultural production and to increase its irrigated acreage, it would risk antagonizing more powerful Egypt and violating the international agreement.

Ethiopia, whose area includes the headwaters of the Blue Nile, has on occasion considered building dams on that river that would reduce flows or change the timing of flows to Egypt. In 1978, then President Anwar Sadat said that Egypt would go to war over the Nile if Ethiopia attempted to reduce flow. Other Egyptian officials, including the previous foreign minister, have repeated this threat periodically.

Apart from rising demand for water fueled by population growth, other phenomena can affect shared water in ways that increase tensions. Climate change can alter flows in unpredictable ways; toxic contamination can affect shared water supplies, including ground water aquifers; and a variety of strategic issues may affect international relationships over water resources. A disturbing example of this latter influence on shared water resources unfolded in 1990 and 1991.

After Iraq invaded Kuwait, cutting off the flow of the Euphrates River to Iraq was discussed as a matter of military strategy. As the downstream nation on both the Tigris and Euphrates Rivers, the desert nation of Iraq is in a vulnerable position. The upstream nations are Syria and Turkey, never particularly strong friends of Iraq--or of each other.

Some months before Iraq's invasion of Kuwait, Turkey completed the Ataturk Dam on the Euphrates. That dam is part of a large-scale development project involving many other dams and an ambitious expansion of Turkish irrigation. After warning Iraq and Syria of its intentions, Turkey closed the dam and effectively shut off the flow of the Euphrates to Iraq and Syria for one month as the reservoir behind the dam filled.

Iraq and Syria objected, but President Oza of Turkey responded with references to Turkey's right and ability to control the flow of a river running through its territory. In fact, Turkey was probably interested in a point of leverage over Syria, which had been supporting Kurds in southern Turkey. President Oza threatened Syria that if it were to continue to provide support for the Kurds, Turkey could cut off Syria's water supply from the Euphrates. Syria protested to the United Nations, Turkey withdrew the threat, and now Turkey denies that it would use water as a weapon. The reality of Turkey's strength as the upstream nation on the Euphrates with the Ataturk Dam in place is nevertheless plain for all to see (Gleick 1990).

As the United States' most recent international military undertaking, the Persian Gulf war may be a harbinger of strategic issues of the future. In addition to discussions of cutting off Iraq's Euphrates water, the Gulf war brought the intentional destruction of desalination plants, the use of oil fires as a weapon, and the targeting of other water and energy resources facilities for destruction. The Gulf war suggests that we may see more intentional use of the environment and its resources not only as targets but as a tool of war. Given this development, water supplies are likely to be increasingly at risk.

Conclusion

Water is indeed miraculous stuff. It is not really possible to improve it, but it is easy to make it much worse. We can contaminate it, waste it, and do violence getting it or keeping it from others.

The issue of sustainable development and the hope that we can provide a decent quality of life for the 5.4 billion people alive today, not to mention the 12 or 13 billion people we may have in the future, requires that we resolve fundamental questions about long-term water availability and quality. Global climate change and other major environmental threats will complicate that task. Uncertainty about water availability grows; the more we learn of possible global environmental changes, the harder it is to be sanguine about their implications.

It has been said that the two most difficult things to keep out of water are salt and politics; we might add oil to that list. Indeed, we are not keeping salt or other contaminants out of our limited water supplies, and we have not satisfactorily worked out political solutions to the problems of competition and conflict over water resources. Unless we do a better job, the 1990s and beyond are likely to see a series of crises over water resources, crises with potentially ever more grave consequences.

References

Christmas, J., and C. de Rooy. 1991. The decade and beyond at a glance. *Water International,* 16, 127-34.

Food and Agriculture Organization. 1990. *FAO Production Yearbook 1990.* FAO Statistics Series No. 99, vol 44. Rome: Food and Agriculture Organization of the United Nations.

Gleick, P. H. 1990. Environment, resources, and international security and politics. In *Science and International Security: Responding to a Changing World,* ed. E. H. Arnett. Washington: American Association for the Advancement of Science, 501-23.

———. 1992a. Effects of climate change on shared fresh water resources. In *The Challenge of Responsible Development and a Warming World,* ed. I. M. Mintzer. Cambridge: Cambridge University Press.

———. 1992b. Water and energy. In *Water in Crises: A Guide to the World's Fresh Water Resource,* ed. P. H. Gleick. New York: Oxford University Press.

Postel, S. 1989. Water for agriculture: Facing the limits. *Worldwatch Paper 93.* Washington: Worldwatch Institute.

Shiklomanov, I. A. 1992. World fresh water resources. In *Water in Crises: A Guide to the World's Fresh Water Resource,* ed. P. A. Gleick. New York: Oxford Press.

Waggoner, P., ed. 1990. *Climate Change and U.S. Water Resources.* New York: John Wiley and Sons.

World Commission on Environment and Development. 1988. *Our Common Future.* (The Bruntland Commission). New York: Oxford University Press.

World Resources Institute. 1990. *World Resources 1990-91.* New York: Oxford University Press.

9

Tropical Forests and Human Society

James D. Nations

Imagine this scene in modern Guatemala: a Maya Indian family, eating tortillas, sitting on the grass outside a McDonald's fast-food restaurant in Guatemala City. Meanwhile, middle class Guatemalans are driving through the *autoservicio rapido* to ask for *hamburguesas* and *papas fritas* at the drive-in window.

The interior of the restaurant is tastefully decorated in a Mayan motif: plaster archeological pieces above the tables and, on the walls, giant blow-up photographs of a Mayan codex. A codex is a painted accordion-fold book made of tree bark, used as a repository of ancient Mayan knowledge. The ancient Maya stored ten centuries of research and experience on astronomy, mathematics, food crop production, water control, and wildlife management in their codices. These books held the detailed record of a civilization that had flourished in the tropical forest of Mexico and Central America for more than a thousand years. At their peak in the tenth century, the ancient Maya had achieved advances in writing, mathematics, and astronomy surpassed by no other civilization in the world.

To see a real Mayan codex today, one must travel to Dresden, Madrid, or Paris. None exists in Guatemala--for a simple reason. In 1562, a group of Spanish friars led by Bishop Diego de Landa gathered every sacred book they could find in the Mayan world. Here are Landa's own words on what happened:

> These people. . . made use of certain characters or letters, with which they wrote in their books their ancient affairs and their sciences. We found a great number of books in these characters, and as they con-

tained nothing in which there was not to be seen superstition and lies of the devil, we burned them all (Landa [1566] 1978).

Landa and his friars piled the Mayan codices at the foot of a giant ceiba tree in the town of Maní, Yucatan, and torched the books and the tree together. Landa later noted that the friars were amused to see the Mayan priests, forced to watch the event, gnashing their teeth and pulling out their hair in anguish.

The only Mayan codices surviving today are the three or four that conquistadors had earlier sent back to Europe as curiosities. Those curiosities are now prized museum pieces. Scholars who study them find tantalizing clues to the knowledge lost in the bishop's bonfire that day in 1562.

Today, more than 400 years after the Mayan codices were torched in Yucatan, our generation is witnessing a similar event taking place on a much larger scale and with far more serious consequences. This event is the destruction of the biological diversity of our planet. Each living species of plant and animal is a page in a priceless book that cannot be recreated once it is sent up in smoke. Tropical forests are the most intense expression of that biological diversity; they are estimated to contain at least half, and perhaps two thirds--some say nine-tenths--of all species on earth. And those forests are being destroyed at a rate of twenty million hectares annually (WRI 1990, 105-7).

Many citizens, biologists, and conservationists today are not willing to follow the model of the Mayan priests, pulling out their hair in anguish while watching fire destroy the unfulfilled potential of the tropical forest's biodiversity. And they are not willing to let the modern counterpart of the Spanish priests watch in amusement. Thus a race is underway to slow and stop this destruction before these forests are completely consumed. The people striving to slow the forest destruction have recognized the value of tropical forests and their diversity--the range of goods and services that these forests provide today and are likely to provide in the future. Furthermore, there is evidence that the destruction is unnecessary, even to meet current needs. Sustained use of the tropical forests for crops such as spices, nuts, and palm fronds makes the forests more valuable than has been previously recognized. Thus, it is in the self interest of forest users, and all of us, to preserve rather than destroy these tropical forests.

To understand the nature of this race to save the tropical forests, it is necessary first to examine the mechanisms involved in tropical forest destruction.

Mechanisms of Tropical Forest Destruction

Tropical forests occur around the world at low latitudes where temperatures and rainfall are sufficiently high. A map of such forests would show large areas in central Africa, Southeast Asia, and Latin America. In each of these areas, part of the forest is destroyed each year. While each area has its particular pattern of destruction, it is widely accepted that the destruction reaches its peak in Latin America (WRI 1990, 102-3).

Throughout Latin America, tropical forest clearing typically takes place in three stages: road construction, colonization, and cash crop production.

In the first stage, bulldozers clear roads through the forest to enable the harvest of hardwood timber, oil exploration, or the establishment of military control over border areas. Most road construction through tropical forests in Latin America is driven by a desire to log hardwoods that will end up on international markets in the form of office paneling, boat accessories, or fine furniture. Other roads are bulldozed through the tropical forest to look for oil, often in areas inhabited only by indigenous people. In the Amazon rain forest of Ecuador, oil companies from the United States, France, Brazil, and Argentina are exploring for oil inside Ecuador's largest national park, called Yasuní, which is also the home of the last uncontacted Waorani Indians.

No matter why roads are built through the tropical forest, they almost always introduce the second stage of forest destruction, colonization. During this stage, farm families from other regions of the country file down the new roads and filter into the forest to clear new land for slash-and-burn agriculture. Families are driven into the tropical forest in part because the most fertile lands in Latin America are owned by the small percentage of the population that is wealthy. This group frequently uses the land to produce cash crops for export rather than to produce food for the local population. The bulk of a region's rural families are left with the need to produce food on the remaining, often less productive, land. Lowland rain forests, many of them inhabited by indigenous groups, are an easy destination for Latin American migrant farmers, who clear the vegetation in hopes of growing crops. Unfortunately, the tropical rain forest soils give them only meager yields for a couple of years, after which it becomes necessary for the migrants to clear more forest if they are to continue farming.

Another reason these families follow roads into the forest is that most of the jobs available to them in the cities offer no more pay or chance for advancement than selling bars of soap at stop lights. There are also too few full-time jobs on the land, because large-scale farms usually require outside labor--for their export crops--only at harvest time. And some export crops, such as beef, use so little labor that they actually push families off the land.

A final aspect of the picture is that there are more and more people in this situation each year. Today, more than 460 million women, men, and children live in Latin America. If current growth rates do not fall, that number is expected to double within 34 years (PRB 1991). Given growth on this scale, one can expect landless families to migrate to forests as long as any tropical forests remain to be colonized.

The pioneer farm families who move into the tropical forest are often followed by the third stage of tropical deforestation, the expansion of cash crops into the forest. During this third stage, land cleared by colonists is taken over by a second wave of settlers who buy up the farmers' clearings and use them to produce crops like pineapples, bananas, or beef cattle.

Because beef cattle production results in such a meager yield per unit of land, it is the most wasteful and most destructive of these third stage cash crops. Most of the cattlemen who follow the farmers into the forest are less interested in producing beef than in gaining control of the land as a hedge against inflation. In highly inflationary economies like most in Latin America, land is one of the few possessions that maintains its value. But despite the predominant interest in land speculation, cattle are usually put on the land just to show that it is being used. Throughout Latin America having cattle on the land prevents government officials from taxing it as unused and prevents new colonists from squatting on the land and trying to claim it.

Some of the beef that is produced on cleared forest land in Latin America is exported to the United States, where North Americans eat it in luncheon meats and hamburgers. But most of it ends up in the Latin American capital cities, to be eaten by the urban middle class as they stop at the drive-through windows at the local fast-food chain.

Meanwhile, the Mayan family sits on the grass, eating their last tortilla, contemplating a move to the tropical forest as their only hope to produce food for next year.

The Value of Tropical Forests and Biological Diversity

Why is it important to pull the genetic books of tropical forest biodiversity out of the fire to protect them as the Mayan codices were not?

The reason is clear, though many of us are not aware of it: tropical forests and the biodiversity they hold have been vitally important to human civilization in the past, and they are likely to be at least as important in the future.

If you have eaten rice, bananas, pineapples, or avocados, you have eaten crops that originated in the tropical forest. A short list of tropical forest crops includes sweet potatoes, manioc, sugarcane, oranges, papayas, mangos, peanuts, cinnamon, cloves, and coffee. Without tropical rain forests, the Swiss would have no chocolate, the French would have no vanilla, and the Italians would have no tomato sauce. These three crops originated in tropical forests--in fact, in the tropical forests of Latin America. Tea originated in the tropical forests of southeast Asia, as did sugarcane, bananas, and oranges.

Cacao beans for chocolate originally came from the Amazon rain forest. Its wild precursors still grow there, and plant breeders use those wild strains to modify domesticated varieties to improve their taste, or productivity, or resistance to disease. Indeed, the germplasm of the tropical forests is critical to the continued health of existing food crops, most of them more central to the human diet than chocolate.

It has often been noted that the world depends for food on a small group of plants (Oldfield 1984, chap. 2). Indeed, 15 plants stand between humanity and starvation: five cereals, three root crops, three seed legumes, two sugar crops, and two tree crops supply the majority of human food. Those crops are rice, wheat, corn, sorghum, barley, potatoes, manioc, sweet potatoes, beans, soybeans, peanuts, sugar cane, sugar beets, bananas, and coconuts. In addition to the smallness of this list, most crops on the list have relatively narrow genetic diversity. Thus, a single epidemic disease can threaten the species and the people who depend on it.

A mosaic virus once threatened to wipe out manioc in Africa because the varieties cultivated were based on a narrow range of genetic diversity. Manioc is one of the three root crops on which humans depend and is one of several crops that emerged from the tropics of the New World and is now a staple in the Old World. Americans eat it mainly in the form of tapioca pudding, but it is the daily bread of 300 million people in other parts of the world. People living in Africa did not lose that daily bread to the virus only because of the diversity in the manioc species in the Amazon region. Plant geneticists found wild strains and traditional varieties grown by indigenous tribes in the Amazon that they were able to cross-breed for greater resistance to the mosaic virus and so save manioc as Africa's staple carbohydrate (Cock 1979; Hahn and Keyser 1985).

Seeds stored in gene banks can conserve some food crop varieties, but gene banks are vulnerable--especially in the tropical world--to accidents, power failures, and destruction during civil wars. More than that, species in cold storage cease to evolve, while their natural cousins continue to evolve with the weeds and diseases and insect pests around them. Only standing, healthy tropical forests provide the species diversity safety net required for resilient crops. In our attempt to understand and use the species diversity that tropical forests contain, our natural allies are the indigenous people who maintain traditional ways of life. All of the 15 major food crops came to us from traditional people, and their descendants preserve the indigenous cultivated varieties today. Called landraces by agronomists, these indigenous varieties play an important role in maintaining germplasm for crop development throughout the world.

Some Mayan farmers in the highlands of Guatemala--cousins of the family sitting outside the McDonald's--have races of corn that will grow on one side of the hill but not on the other, races that grow best in dry years, and others that grow best in wet years. With a dozen races of corn in a field, a farmer always has one or more that will produce in any given year. The wild and primitive germplasm growing in the forests and fields of indigenous people is the best insurance we have against the destruction of the human food supply.

Healthy tropical forests are also a form of insurance that new food crops will become available in the future. Skepticism that no new foods are likely to be introduced is misplaced, as shown by the recent history of new fruits in the United States. A century ago, bananas were an exotic food in the western world. When first introduced into the United States, they were sold individually wrapped for two dollars each. Now bananas are one of our most common fruits. Kiwi fruits were met with skepticism when they were introduced in 1962, but by 1988, 10 million Americans were eating them. The U.S. avocado industry owes its existence to Wilson Popenoe's trip to Guatemala and Mexico 70 years ago, where he gathered 26 kinds of avocado seeds growing in the forest and in the household gardens of Maya Indians (Smith 1985, 30).

Incidentally, when first introduced into the United States, avocados were viewed as so strange that few would eat them. In response, their promoters placed advertisements in all the major newspapers fervently denying that avocados were a strong aphrodisiac. Sales immediately skyrocketed.

Twenty years ago, the average U.S. supermarket carried 65 kinds of fruits and vegetables. Today, many of them have at least 140 species of fruits and vegetables, and some of them have 250. One supermarket in

Florida that has better than average access to the produce of Latin America's tropical forests sells 400 different fruits and vegetables during the course of the year. As tropical forests have provided new foods in the past and right up to the present, this home of at least half of all the species on earth is likely to provide new crops in the future.

Tropical forests have also provided non-food materials such as rubber, gum arabic for ink, and the glue for postage stamps. As with food crops, Europeans learned about the plants that produce these materials by observing how indigenous peoples use tropical forest plants. We have learned much from the Indians of Central and South America, the pygmies of Africa, and the traditional tribal people of Asia.

Curare was discovered by Indians of the Amazon as a poison for the tips of blowgun darts. Today, injected into people during open heart surgery, it freezes the movement of the heart and lungs so the surgeon can operate without dodging the normal motions of body parts. Indians in the Amazon use the barbasco vine to stun fish and make them easy to collect. An American chemist observed this use of the vine and analyzed the plant, finding alkaloids that led to the development of cortisone for treating skin diseases, and to the development of birth control pills. Seventy percent of the plants known to have anti-cancer properties come from tropical forests. And most of them were introduced to us by traditional peoples who live in the tropical forest themselves.

Sustainable Use of the Forest

There is more to learn from traditional peoples--and others--occupying tropical forests. They can teach us that a continued, sustainable use of the forest provides more value than does cutting it down for cattle farms or land speculation.

The Mayan inhabitants of the Guatemalan Peten were pushed out or killed during the Spanish conquest of the sixteenth century. During the late 1800s, Spanish-speaking immigrants moved in to harvest chicle trees in the Peten's tropical forest. Their descendants are still there today, producing chicle, allspice, and Chamadoria (xate) palms in a harvesting system that threatens neither the individual plants nor the forest itself. Combined, the three products--chicle, allspice, and palms--produce $7 million per year in export revenues for Guatemala and give the 6,000 families who make a living from this harvesting a reason to protect the tropical forest.

The families harvest and export 100 million fronds of Chamadoria palms from the Peten every year. These palm leaves are used in Europe and the United States in floral arrangements. Xate palm is also cut and provides the "green screen" for cut flower arrangements. Xateros--people who harvest the xate palm foliage--walk through the forest clipping leaves from palms and placing them in bags they carry over their shoulders. Cutters return to the same palm every three months and clip off the new leaves that have grown since the last visit. Harvesting does not harm the palm or the forest, yet it produces income for the harvester and for Guatemala as a whole.

Successful sustainable use of the forest such as palm harvesting suggests that the forest may be worth more standing than cut, worth more alive, as a source of renewable products, than it is cleared, burned, and turned into short-lived corn fields and cattle ranches. Harvesters obviously have an economic interest in forests, but, according to this evidence, so does Guatemala as a whole.

One sustainable use sometimes promotes another. Protecting the forest for renewable forest products also protects it for tourism. In the Peten, tourism now brings in $22 million each year. Such an argument gives Guatemalan politicians an additional reason to protect the forest against loggers and cattle ranchers.

In Esmeraldas Province of western Ecuador, Conservation International is working on a similar project with a community of coastal blacks. This community harvests vegetable ivory--tagua--from a tree that grows wild in the tropical forest. Before World War II, 30 percent of all buttons were made from slices of tagua nuts. Then plastics became cheaper, and tagua was dropped. But now, people are willing to pay extra to use a product that not only does not harm the environment, but actually helps protect it. Licensing arrangements have been made with clothing companies Patagonia and Smith and Hawkin to use tagua buttons from Ecuador on the clothing in their stores and catalogues. Money that comes from these licensing arrangements returns to the community to demonstrate that forests are more valuable alive than dead.

These examples of economically-viable sustained uses of tropical forests stand in sharp contrast to arguments based only on intrinsic or spiritual values for these forests. Economic arguments are essential: the combination of rapidly growing populations in the tropical regions and the fragile nature of national economies there doom to failure any argument that we should protect tropical forests simply because non-human species have a right to exist.

Stephen Jay Gould makes this same argument in a recent article (Gould 1990). Our focus as a species and our focus as conservationists, he writes, should be one of enlightened self-interest. We should not try to conserve biodiversity for some distant planetary future, but for our own existence today. We have a legitimate interest in our own lives, he says, in the happiness and prosperity of our children, and in the suffering of our fellows, indigenous peoples and colonists included.

The difficulty arises in just what "enlightened" self interest is, particularly in the time scale over which the happiness and survival of humans is to unfold. The scholar observing a farmer attempting to grow food on a recently deforested area can make certain forecasts. The farmer's children are likely to be poorer in the future than the farmer is today, lacking forests to serve as individual or national resources. The scholar can add that the world will also be deprived of needed genetic resources. The farmer will respond that he need not worry about the children's adult future if they are left to starve to death next year.

In this chapter, the emphasis has therefore been on those steps that provide a livelihood *in* the forests so that a transition can be designed to a longer-range view while the forests are protected by their productivity. The key to the future of tropical forests and biodiversity may be "use it or lose it." But we need to find ways, such as harvesting chicle, xate, and tagua, and promoting tourism, to use forests without destroying them. And we need to make clear that we are not talking about conserving tropical forests and biodiversity out of an altruist vision of protecting the earth for other species, but because people, now and in the next generation, need the genetic diversity stored there.

That same self interest is the concept that has to drive attempts among industrial countries to conserve tropical forests and biodiversity in the Third World. The more prosperous countries have the luxury, denied to the indigenous peoples of the tropics, of taking a longer view. Thus, we must add to our direct concerns for the future of human food supplies, medicinals, and other materials our concern for the contribution that deforestation makes to human-induced global climate changes. Those changes can shift the climate in ways troublesome to all sectors of society (see Firor, this volume).

Conclusions

This chapter has described the forces that are leading to the rapid destruction of tropical forests in Latin America, and it has suggested the kinds of arrangements that can protect these areas as we move into the critical decades for the future of tropical forests, biodiversity, and indigenous people. The central element in those arrangements is the demonstration to government leaders, bankers, and rural families that the advantages of conserving biodiversity outweigh the advantages of destroying that diversity for short-lived profits and immediate needs.

We have a long way to go, but by working to protect tropical forests and biological diversity, we can assure their survival and the survival of the people who depend on them. Ultimately, that includes not just indigenous people, but all of us.

References

Cock, J. H. 1979. Cassava research. *Field Crop Research* 2, 185-91.
Gould, S. J. 1990. The golden rule: A proper scale for our environmental crises. *Natural History.* September 1990, 24-30.
Hahn, S. K., and J. Keyser. 1985. Cassava: A basic food of Africa. *Outlook on Agriculture* 4(2), 95-99.
Landa, Friar Diego de 1978 [1566]. *Relación de las Cosas de Yucatán*, published as *Yucatan Before and After the Conquest.* Translated by William Gates. New York: Dover Publications.
Oldfield, M. L. 1984. *The Value of Conserving Genetic Resources.* Washington, D. C.: U.S. Department of the Interior, National Park Service.
Population Reference Bureau. 1991. *World Population Data Sheet.* Washington, D.C.
Smith, N. J. H. 1985. *Botanical Gardens and Germplasm Conservation.* Honolulu: University of Hawaii Press.
World Resources Institute. 1990. *World Resources 1990-1991.* New York and Oxford: Oxford University Press.

PART FIVE

Designing the Future: Coping with the Crises

10

Creating an International Process to Address Greenhouse Warming

William A. Nitze

The Nature of the Challenge

Greenhouse warming and associated climate change present humankind with its most far reaching environmental challenge to date. Scientific consensus now exists that emissions from human activities are substantially increasing atmospheric concentrations of greenhouse gases (carbon dioxide [CO_2], methane, chlorofluorocarbons [CFCs], and nitrous oxide) and that these increases will enhance the natural greenhouse effect, resulting on average in an additional warming of the earth's surface (see Firor, this volume). The Science Working Group of the Intergovernmental Panel on Climate Change (IPCC) has made a mid-range prediction that under a business-as-usual scenario, global mean temperature is likely to increase about one degree centigrade above its present level by 2025 and about three degrees centigrade before the end of the next century (IPCC 1990).

Greenhouse warming of this magnitude would have varying effects on different countries and regions depending on their geographic location, economic and social structure, and level of development. It is fair to say, however, that most countries would experience some negative effects from this rate of change, that poorer countries would suffer more than richer ones, and that natural systems, agriculture, and forestry would be more adversely affected than less climate-dependent sectors of the economy in all countries. There is a significant risk, moreover, that the world's climate will change in unexpected and potentially catastrophic ways. This risk will increase geometrically over the longer term if human-induced emissions of CO_2 and other greenhouse gases are allowed to increase at currently projected rates.

We already have the technology to reduce significantly human emissions of CO_2 and other greenhouse gases and to improve our ability to adapt to that climate change to which emissions to date have committed us to at little or no cost to the world's economy. The National Research Council has published a report that identifies a wide range of cost-effective measures to reduce or offset greenhouse gas emissions, particularly in the energy sector (National Research Council 1991). It is estimated in that report that collectively these measures could reduce U.S. greenhouse gas emissions by 10 to 40 percent of 1990 levels at low cost. Other studies, including a report by the Office of Technology Assessment, have found an even greater potential for emissions reductions from energy efficiency improvements, afforestation and other cost-effective steps (OTA 1991).

Taking these and similar steps globally would not only provide low-cost insurance against the risks of greenhouse warming but they are seen by many as essential to providing the world's growing population with a decent quality of life. The countries of the former Soviet Union, China, India, Brazil and Eastern European and developing countries are already facing severe capital constraints in increasing their food, energy, and industrial production at a rate sufficient to meet the expectations of their people (Levine et al. 1991, 33-35). Even in the unlikely event that they could find the trillions of dollars required to clear more land, produce more agricultural chemicals, build more power stations, and make more steel, the environmental costs of such a traditional development strategy would soon become intolerable. The best hope for these countries to obtain a reasonable quality of life for their people is to decouple developing their economies from increasing their consumption of fossil fuels and depletion of natural resources. The United States and other richer countries in turn are faced with the responsibility of helping the poorer countries reach this objective if we are to pass an acceptable environment on to our children and grandchildren.

For the world's countries to cooperate in reducing the risks of greenhouse warming, a fully international process is required. This process should combine "top-down" and "bottom-up" negotiating procedures. From the top down, international bodies and organizations can establish principles and goals that governments set more specific objectives to meet. Governments can also create incentives so that private firms, industrial organizations, environmental organizations, local governments and citizens groups can, from the bottom up, develop and implement even more specific strategies to reduce emissions and build adaptive capacity. Specifically, central governments can work with private sector organizations to provide information about available technologies and practices, reduce subsidies on consumption of energy, water and other natural resources and provide fiscal incentives for their efficient use, and support local decision-making by improving education, training, communications, and basic infrastructure.

Finally, the process should allow continual evolution of goals, targets, implementation mechanisms and institutional arrangements as we learn more about the science and probable impacts of greenhouse warming and the effectiveness of various policy responses (Nitze 1990).

The 1987 Montreal Protocol on Substances that Deplete the Ozone Layer (UNEP 1987) provides a precedent for the type of iterative process required. That international agreement established initial targets for reducing or stabilizing production and consumption of CFCs and some other ozone depleting chemicals. It provided for periodic review of those targets in light of new knowledge generated by an continuing inquiry into the scientific evidence of ozone depletion, the effects of CFC emissions, the availability of substitutes for CFCs and other ozone-depleting chemicals, and the costs of shifting to those substitutes over various time scales. Developing countries were given a 10-year grace period before they had to meet the targets. The Protocol also contained provisions with respect to emissions reporting, technology transfer, trade sanctions, and adjustments to certain provisions. The Protocol was amended by the signatories in June 1990 to require the complete phaseout of most ozone depleting chemicals by the turn of the century and to create a special fund to assist developing countries in making the transition to non-ozone depleting substitutes.

Each of these elements could be replicated in the climate change convention now being negotiated by the Intergovernmental Negotiating Committee for a Framework Convention on Climate Change and in subsequent protocols to that convention. The convention could establish initial targets for stabilizing and subsequently reducing emissions of greenhouse gases other than ozone-depleting chemicals controlled under the Montreal Protocol, particularly in the United States, Europe, and Japan. The IPCC process for assessing the science and social and economic impacts of climate change and evaluating possible policy responses could be formalized and continued under the convention. The convention could also establish separate emissions reduction obligations for developing countries, reflecting their special circumstances, and provide for technology transfer and financial assistance to those countries in meeting their obligations. The convention could provide for subsequent amendments and protocols in light of information provided by the assessment process. Finally, the convention could involve innovative provisions on data gathering and analysis, emissions reporting, and non-compliance.

The greater complexity of the greenhouse warming issue may require the convention to provide for a bottom-up policy process that goes beyond the Montreal Protocol in two critical respects. First, the convention could require each party to prepare its own national strategy for addressing greenhouse warming and to share that strategy with other parties. (The Montreal Protocol does not contain such a requirement, since it deals with a single class of chemicals produced by a small group of manufacturers concentrated in the United States, Europe, and Japan.)

Any international strategy for reducing greenhouse emissions will necessarily take into account a wide range of sources and sinks of CO_2, methane, and other gases, the relative importance of which differs substantially from country to country. The most cost-effective approach to reducing net greenhouse gases will similarly differ from country to country, depending on economic and social circumstances and the existing mix of public policies. Specific national strategies could be coordinated and harmonized in subsequent phases of the international negotiating process.

Second, international organizations, local communities, and non-governmental organizations should be more involved in developing and implementing national and international strategies than they were with ozone. The Montreal Protocol took several important steps in this direction. The United Nations Environment Program took the lead in negotiating the protocol and has an ongoing role in monitoring compliance with it. The World Bank is playing an important role in providing assistance to developing countries in replacing CFCs with substitutes that do not deplete ozone. The Natural Resources Defense Council and other environmental organizations put pressure on the U.S. Environmental Protection Agency and American producers of CFCs to support a strong international agreement in lieu of controls applicable only to American companies. Du Pont and other major producers mounted a major effort to accelerate development and commercialization of substitutes for the most ozone depleting CFCs.

Any international regime to address climate change will have to go much further in this direction. Cost-effective and politically acceptable strategies for addressing global warming are likely to differ not only among countries but among different regions and local communities in the same country. A rational transportation policy for Wyoming will not be the same as one for New York. Optimum agricultural practices in Southern Brazil will differ greatly from those in the Amazon basin. Land use policies in low-lying coastal areas will not resemble those in central highlands. Strategies so diverse and complex require the input of local interest groups and other local people affected by global warming and by the steps to diminish it. Appropriate local practices such as cultivation of native forest products, multiple species cropping, integrated pest management, and drip irrigation are not easily legislated from the center.

Such a system for negotiating an international regime for reducing global warming will not occur in a vacuum. It can be facilitated by--and in fact needs--simultaneous changes in technology, economics, and politics.

The New Technology

It is hard to imagine improving the quality of life for the world's rapidly growing population of poor people without decoupling economic development from increasing consumption of fossil fuels and depletion of natural resources. Since the world's current population of 5.4 billion people is already generating an unacceptable level of pollution and natural resource depletion, our goal should be to offset increases in world economic activity and population growth with even greater decreases in world pollution and resource depletion. If world population were to grow at 1.5 percent annually and average per capita GNP at 3.5 percent per year, the rate of decrease in per capita pollution and resource depletion would have to exceed 5 percent per year to put us on a sustainable path. This rate of change may appear ambitious, but the new technology available to achieve it is already emerging.

The world is undergoing at least four interlocking technological revolutions. The first is a shift from resource-intensive to knowledge-intensive production. The United States, Europe, and Japan are already generating the bulk of the value added in their economies by the manipulation of information and the provision of services. Although these activities require a certain level of energy and other resources for offices, telecommunications networks, computers and transportation, they are much less resource-intensive than traditional manufacturing.

This trend has in itself substantially reduced consumption of energy and raw materials per unit of GNP, in addition to the reduction attributable to technological improvements in energy and manufacturing systems. Delivered energy use in the industrial sector of the U.S. economy, for example, declined from 26 quads in 1973 to just over 20 quads in 1986.

If trends underway in 1972 had continued, U.S. industry would have used nearly 32 quads in 1986. Of the nearly 12 quads of delivered energy saved, nearly half was attributable to a change in the composition of U.S. industry (Department of Energy 1989). This trend, observable in Europe and Japan as well, has in turn contributed to a reduction in the volume of greenhouse gases and other pollutants emitted per unit of GNP in advanced economies.

The second technological revolution involves continuing improvement in energy and manufacturing systems themselves. Technological innovations in individual production processes such as steel making or ethylene production are responsible for part of the change. These innovations have steadily reduced the amount of energy and raw materials required to make a ton of steel, ethylene or other final product. Another part of the improvement comes from the statistical quality control revolution in manufacturing itself. Developed by American statistician W. Edwards Deming during the Second World War, statistical quality control techniques enable manufacturers to increase the efficiency of their manufacturing process dramatically by improving product quality and reducing defects.

One of the most impressive results of the revolution in manufacturing technology is the improvement in energy efficiency. From 1971 to 1986 the energy intensity of industrial processes in the United States--the amount of energy required to produce a ton of steel or a kilogram of polyethylene--has declined by between 1.5 and 2 percent a year. These efficiency improvements, combined with the shift towards less resource intensive production, have led to an annual decline of 1 percent in total energy use despite 2 percent annual growth in manufactured output (Ross and Steinmayer 1990, 89). Similar improvements have been realized in other industrial countries. There is no reason to believe that these improvements will not continue as companies optimize the cost of existing processes, introduce process refinements, and develop entirely new manufacturing methods, such as the elimination of coke ovens through direct steel making. The buildings and transportation sectors also have the potential for significant efficiency improvements through the introduction of better insulation, compact fluorescent lighting, more efficient heating and air conditioning equipment and appliances, and low emissivity windows, as well as modifications in automobiles and other vehicles.

The third interlocking revolution is the explosion in information technology. The ability to collect and manipulate geometrically increasing amounts of information on smaller and smaller pieces of silicon, gallium arsenide and similar materials has given us a tool for optimizing technologies, processes, and ways of using them unimaginable a generation ago. Opportunities for using information technology to improve efficiency are almost limitless. In manufacturing, computer-aided design, computer-aided manufacturing, robotics and other manifestations of the microelectronic revolution are already producing improvements in industrial productivity, product quality, and environmental performance. When sophisticated computers are linked with sensors that monitor the conditions inside a reactor vessel, a column or a waste stream, and with control devices such as pumps, valves or power supplies, they can precisely control the flow of material or energy inputs. Energy and materials are saved by applying only the required amount of an input and only when needed. The quality and uniformity of the product are also improved, resulting in fewer discards and rejects (Heaton, Repetto, and Sobin 1991, 18). Information technology has the potential to produce similar changes in agriculture. By enabling farmers to optimize crop rotation, irrigation, fertilization and pest control, computer-aided agriculture could simultaneously improve yields, reduce resource inputs and control environmental pollution.

The fourth revolution is the development of biotechnology. In the last 20 years, scientists have made extraordinary progress in translating growing knowledge of the genetic code into technologies for manipulating the characteristics of specific organisms. By permitting the more rapid and precise transfer of genetic traits into agriculturally useful organisms than do conventional breeding mechanisms, biotechnology could enable us to develop new strains of plants and animals that are disease- and pest-resistant and plants that are tolerant of drought and other resources scarcities. Biotechnology also has many potential applications in improving industrial production processes and preventing pollution. The bacterium *Thiobacillus ferrodoxins* is already playing a critical role in 30 percent of U.S. copper production and is credited with saving the domestic copper industry (Heaton, Repetto, and Sobin 1991, 18). Newly developed bacteria that break down liquid organics and other hazardous substances have the potential to reduce the cost and environmental risks of hazardous waste disposal. There are potential risks from the release of genetically modified organisms into the environment, but in the view of many, these risks are far outweighed by the potential benefits of biotechnology.

The real challenge is not in developing the "hard" technology, but in developing the "soft" technology for ensuring that existing hard technology uses resources more efficiently and cleanly. This challenge exists in the United States and other developed countries, where there is a large gap between best practice and average practice in most sectors of the economy. The gap is much greater in Eastern Europe and developing countries, where reliance on outdated technologies is compounded by lack of training and wasteful operating practices. Developing and deploying the soft technology necessary to narrow the gap between the potential and the actual, particularly in developing countries, will require fundamental changes in economic incentives, an unprecedented level of international cooperation, and a restructuring of political systems at the top and the bottom. Thus agreement to lessen global warming is also agreement to lessen some of the world's most pressing economic and political problems.

The New Economics

Current methods of measuring income at the national, corporate or individual level do not reflect the environmental or social welfare impacts of the activities generating that income. A country can increase its gross national product (GNP) by depleting its natural resources to produce goods that are not marketable internationally while severely polluting its soil, water and air. This is precisely what has been happening in Eastern Europe since the Second World War and is happening today in many developing countries. Even in the United States, where considerable effort has been made to control pollution through environmental regulation, our national income accounts do not reflect the long-term cost of removing the phosphates, nitrates, heavy metals, organic compounds and other pollutants that we have been depositing into the environment. It has been estimated that if U.S. GNP between 1950 and 1986 were adjusted to reflect the remediation costs of removing those pollutants, the "adjusted" annual GNP per capita would have increased from $2,500 (in 1972 dollars) in 1950 to only $3,400 in 1986 rather than the unadjusted level of $7,200 (Daly and Cobb 1989, pp. 418-19). In Eastern Europe and many developing countries, adjusted per capita GNP levels would have shown substantial declines over the same period.

The corollary to the need to adjust national income downwards to reflect depletion of natural resources or damage to the environment is the need to adjust national income upward for investments that increase natural resources or reduce pollution. Under conventional accounting methods, for example, the very substantial investments in flue gas desulfurization and other technologies to reduce SO_2 emissions by U.S. utilities over the last 20 years are reflected as costs that produce no corresponding benefits. If the environmental benefits of those investments--reducing acid rain, for example--were included in the calculation, the adjusted rate of return on the investments might be extremely high. Similar adjustments should be made in calculating the benefits of other pollution prevention, natural resource conservation and reforestation investments.

Many people instinctively understand that their actions have environmental impacts not reflected in the conventional accounting of economic activities. But these same people have no reason to give sufficient weight to those impacts in their business decisions when preventing pollution or conserving natural resources involves short-term costs and no benefit captured by conventional accounting. Politicians particularly fall prey to this thinking. They want to measure their success or failure in terms of short-term changes in income or employment. Internalizing environmental externalities into income and balance sheet accounting systematically could motivate public and private decision makers to give sufficient weight to environmental impacts in their decision making.

One way of accomplishing this objective would be to add an environmental category to inputs, outputs and changes in capital stock. Natural resource drawdowns such as loss of wildlife habitat or watershed protection associated with deforestation would be included as environmental inputs or costs. Degradation of air, water or soil quality associated with industrial emissions or other waste streams would be included as negative environmental outputs and netted against the value of the production with which they were associated. Both types of environmental impacts would be reflected in asset accounts as reductions in the value of environmental assets in the same manner as depreciation or depletion of market assets. Conversely, investments in improving natural resources or cleaning up past pollution would be reflected as increases in the value of environmental assets (Zimmerman 1991).

Accounting explicitly for environmental impacts in the manner suggested above, as well as the removal of direct government subsidies for environmentally harmful activities, have important implications for government policy. Political and economic interests in the U.S. and other countries resist an accounting change that highlight their pollution; they resist even more strongly any loss of subsidy. The ability to pollute air, water, or land without consequence represents a subsidy enjoyed by polluters. As soon as a government adopts regulations forbidding pollution, taxes on pollution, or even efficiency or environmental performance standards resources, introducing energy, carbon or pollution taxes, or even to imposing efficiency or environmental performance standards on manufacturers or users of polluting technology or equipment, a subsidy is lost. Enlarging the context for discussion of these subsidy-ending policy options could broaden public support for them. For example, the U.S. currently taxes income and labor much more heavily than it does energy and other natural resources. A shift from income and social security taxes to energy and pollution taxes would provide overall welfare benefits.

It will not be easy to value all natural resources or environmental benefits. Traditional consumer and investment goods have market values; most environmental goods, such as clean air or a view of Grand Canyon, and many natural resources, such as rainfall or today's climate, do not. One method for establishing such values is to make a political determination of the maximum discharge or emission level for a given region or country, based on the desired cleanliness of the air or freedom from induced climate change, and then to allocate discharge or emission levels for particular industries within that maximum level in the form of tradeable permits. The prices at which those permits change hands would then establish a market value for eliminating a unit of the pollutant in question. This approach, which has been incorporated into the acid rain title of the U.S. Clean Air Act, is being considered for reducing the costs of mitigating carbon dioxide and other greenhouse gas emissions (Grubb 1989).

Another valuation technique is to assign arbitrary but conservative values to the environmental costs or benefits associated with particular primary energy sources and then to require the use of those values in selecting the least-cost option for producing a given level of energy services. This approach is being implemented by several state-level regulatory commissions in the United States. A third method is to aggregate the various environmental impacts of producing a kilowatt of electricity from coal or a ton of newsprint from virgin timber. Work on these and other approaches needs to be given high priority.

In some countries it would be necessary to reform basic pricing systems before introducing prices for environmental goods. In many of these countries fossil fuels, electricity, agricultural chemicals and other inputs with environmental externalities are priced below their conventional costs of production. Raising those prices to above conventional production costs would in itself provide a powerful incentive to reduce waste and the pollution associated with it. As a second step, these countries should provide economic incentives for the more efficient use of energy and other resources similar to the incentives being given U.S. utilities to invest in energy efficiency improvements in their customers' facilities. Measures such as these are necessary first steps; more sophisticated instruments for reflecting environmental externalities would come later. And before any of these steps are possible, developing countries will need outside encouragement and assistance.

The New Politics

The world already has many of the hard technologies needed to start decoupling economic growth from resource depletion, emission of greenhouse gases, and other forms of environmental pollution. We have also learned a great deal about the soft technologies required, including methods of industrial and natural resource management, ways of providing incentives for more efficient use of energy and other resources, approaches to internalizing environmental costs and benefits in our economic accounting, and policy instruments for minimizing the costs of environmental protection. The hardest challenge will be to overcome the inertia and resistance to change built into our political systems. A new politics that facilitates change to a more sustainable way of life at the international, national and local levels is the final topic discussed in this chapter. The critical elements of that new politics are local participation, institutional reform, and international cooperation.

The first critical element of the new politics is to involve the local communities, private organizations and individuals directly affected by and affecting climate change and other unsustainable activities. People at the local level need the basic tools and incentives to change their behavior. Education about the environmental consequences of different methods of growing crops, producing other goods and services, providing cooking energy, heat and light to their homes, and transporting themselves from place to place is one step. Training in how to use the methods that are most appropriate for their circumstances is another. Provision of the seeds, energy supplies, transportation facilities, and other basic infrastructure necessary to put those methods into practice is yet another. Finally, the right incentives to minimize pollution and use resources efficiently must be present or the education, training, and infrastructure will go unused. It is difficult to persuade people to use electricity or water efficiently if they are provided free of charge.

In addition, local people need a voice in deciding on projects that affect their livelihood. Much of the environmental damage resulting from the industrialization of Eastern Europe and many developing countries since the World War II has resulted from a failure by all-powerful politicians at the center to inform or consult with local people. Many of the environmentally destructive projects whose consequences we are dealing with today would probably not have gone forward or would have been significantly modified if local people had been informed and consulted. The rise of environmental groups in Eastern Europe and the 1989 revolution have already led to the cancellation of a major water diversion project in the former Soviet Union, the indefinite postponement of the Nagymoros Dam, and the cancellation of a number of nuclear power and other projects. More powerful local participation--such as the Chipko movement in India--could produce similar consequences elsewhere in the developing world.

The second critical element in achieving the changes needed is institutional reform. We already have a panoply of institutions including the United Nations Development Program, the World Bank and overseas development agencies dedicated to achieving needed development. These institutions have not been very successful, however, in bringing about the changes in local practices and behavior necessary to minimize pollution and conserve natural resources. All too often these institutions have supported large projects such as penetration roads, hydroelectric dams, irrigation schemes, and minerals extraction complexes proposed by politicians at the center without sufficient consultation with affected local people or evaluation of environmental impacts. They have been much less successful in supporting the thousands of small-scale local projects necessary to achieve sustainable agriculture, forestry and industry in developing countries. Whatever success they have had has been largely due to the participation of local or international NGOs that are capable of organizing and managing such projects.

This unhappy history has led to a growing recognition that development assistance institutions must give top priority to strengthening human infrastructure at the local level. One way of doing this is to provide funding directly to international and local NGOs for local projects. Another is to support regional and national research centers where local experts can study existing technologies, conduct research, and exchange ideas. A third is to provide funds for education, training and research in developed countries. A common theme in all of these approaches is to move decision making down to the local level to the extent possible.

The third critical component to achieving the kind of world envisioned here is greater international cooperation. Traditional mechanisms for negotiating international agreements should be expanded and streamlined so that countries can reach agreement on environmental issues quickly. If it is cheaper for Germany to clean up its air by investing in Polish scrubbers than in reducing its own SO_2 emissions, it should have a mechanism for doing so. If the United States can more effectively promote economic development in the Caribbean by phasing out its sugar quotas than by providing additional aid, it should eliminate the quotas. If developed countries want Brazil to protect its forests to preserve the global environment, they should compensate Brazil for doing so. All of these transactions require an unprecedented level of international cooperation. We have begun to encourage that type of cooperation through existing international agreements on controlling acid rain and protecting the ozone layer. We will have to go much further in developing mechanisms for truly international strategies to reduce greenhouse gas emissions.

Conclusion

We already have many of the technologies and other tools necessary to put the world on the path required to slow and even halt a greenhouse warming. To use those tools effectively, we must now create an international process for addressing greenhouse warming that integrates the new technology, the new economics and the new politics, all in the context of achieving development that meets the needs and aspirations of the present generation without compromising our ability to met those of future generations.

References

Daly, H. E., and J. B. Cobb. 1989. *For the Common Good.* Boston: Beacon Press.

Grubb, M. 1989. *The Greenhouse Effect: Negotiating Targets.* London: The Royal Institute of International Affairs, 33-40.

Heaton, G., R. Repetto, and R. Sobin. 1991. *Transforming Technology: Agenda for Environmentally Sustainable Growth in the 21st Century.* Washington, D.C.: World Resources Institute, 18.

Intergovernmental Panel on Climate Change. 1990. *Climate Change: The IPCC Scientific Assessment.* Cambridge: Cambridge University Press, xi.

Levine, M. D., ed. 1991. *Energy Efficiency, Developing Nations, and Eastern Europe, A Report to the U.S. Working Group on Global Energy Efficiency.* Berkeley, Calif.: Lawrence Berkeley Laboratory, 33-35.

National Research Council. 1991. *Policy Implications of Greenhouse Warming.* Washington, D.C.: National Academy Press.

Nitze, W. A. 1990. *The Greenhouse Effect: Formulating a Convention.* London: Royal Institute of International Affairs.

Office of Technology Assessment. 1991. *Changing by Degrees: Steps to Reduce Greenhouse Gases.* Washington, D.C.: U.S. Government Printing Office.

Ross, N. H., and D. Steinmayer. 1990. Energy for industry. *Scientific American* 263, 89-98.

United Nations Environment Program. 1987. *Montreal Protocol on Substances that Deplete the Ozone Layer.* Montreal, 16 September 1987.

U.S. Department of Energy. 1989. *Energy Conservation Trends: Understanding the Factors that Affect Conservation Gains in the U.S. Economy.* Washington, D.C.: U.S. Department of Energy. 11.

Zimmerman, M. B. 1991. *Weighing the Costs and Benefits of Climate Change Policies.* Washington, D. C.: The Alliance to Save Energy.

11

Human Impacts

C. M. Hudspeth

It is difficult today to think of a subject more important than the human impact on the environment.

Perhaps the most important statistic about our human impacts is the fact that human population increases by a million in four days. I write this while looking out over Houston, the fourth largest city in the United States. With its surrounding suburbs, Houston is the home of three million people, about the number we add to the earth every twelve days. When one considers the infrastructure necessary to support Houston's population--the energy and food and water--and when one notes the garbage and smog that results, it gives some insight into the impact which the addition of a million people every four days must have on the global environment.

Mercifully, everyone does not live like we live in Houston, but even if Americans lived like most of the rest of the world, it still seems that we would not have achieved a sustainable society.

There is nothing inherently bad about large populations, but the resources which they consume and the uses to which they put modern technology cause degradation in the quality of life and now threaten our survival as a species. We are depleting in decades top soils which nature creates on a time scale of inches per millennium. Forests that have grown over centuries are being cut rapidly, and underground aquifers from the ice age are being rapidly depleted without serious effort to conserve or replace them. Fossil fuels which are formed over geological time spans are being consumed at an accelerating rate. It is estimated that there are five hundred million automobiles in the world today, and if China and India should ever have as many cars per capita as we have in the West, the consequences would be frightening if not devastating.

Seven hundred fishing boats from Japan, South Korean, and Taiwan, equipped with 20 to 40-mile long drift nets, can overnight sweep an area of the ocean the size of Ohio. World fish consumption has more than quadrupled in the last thirty years, and many scientists believe that we have reached or passed the peak of sustainable harvest.

We of course wish to understand how we came to be where we are and to discuss what we can do about it. These are the topics of this book. We reach back to the earliest glimpses, through archeology and paleontology, of possible human impacts on other animal life. We follow these impacts as they expand with the development of irrigated agriculture and crowded cities. And we move through the industrial revolution and on to the pervasive but frequently invisible global impacts of our present age. Finally, we ask what can we do.

A few years ago, Riane Eisler published a book entitled *The Chalice and the Blade* which Ashley Montagu appraised as the most important book since Darwin's *Origin of Species*. While I would not give it that accolade, her interpretation of our history and her prescription for the future are worthy of our attention.

Eisler believes that during a period before written history, societies such as that on Crete treated men and women equally and lived in harmony with nature and such values as love, equality, cooperation, and conservation. Their society was symbolized by the chalice.

Then the first cultural transformation came after metallurgy and the blade. This transformation was not because of metallurgy itself, but the uses to which metals were put. This ancient lesson is applicable today: it is not science and technology, but the uses to which we put them that have contributed to our crises.

With the blade came male supremacy. Kurgan people from the northwest and nomadic tribes who had created a Dominator model of social organization based on power, competition, and male dominance found that they could take by destruction and plunder more easily than they could create. This dominator model which prevailed stood in sharp contrast with the partnership model which it replaced. In Crete the partnership society was not matriarchal but was based on total equality. The ramifications which Eisler sees from this big crossroad in our cultural history are virtually endless. Among other things, according to her, males in the dominator society selected books of the Jewish and Christian Bibles in which girls and concubines, indeed all women, were relegated to an inferior and secondary role (see, also, Redman, this volume, p. 35).

Whom history favors may depend upon the historians. And for 2,000 years the primary values of the dominator society have been taught to our children in the West.

We now face the second big crossroad in our cultural history, she maintains. This occurs because for the first time we now have the technology to destroy ourselves, and we will do so unless we abandon the dominator model in favor of a partnership society that will practice human equality and live by feminine concepts of nurturing and conservation, gentleness, compassion, and peace.

More recently Lester Milbrath published a book entitled *Envisioning a Sustainable Society*. He adopts much of Eisler's interpretation of history and her prescription for the future. While most of us do not try to erect a hierarchy of values by which to live, Milbrath maintains that our two top values should be: First, to preserve the viability of our ecosphere (for without it, what else matters?), and second, to nourish the good functioning of society. Perhaps we could add a third core value from the late Bishop Pike: to actualize the greatest number of human potentialities. Numerous instrumental values must come into play to support these core values. Milbrath envisions a society that would be sustainable, regenerative, flexible, safe, and locally controlled. Modern society, by contrast, gives primacy to power, economic growth, competition, and consumption.

When we look at business enterprises, we tend to look first at the bottom line. As one who has overseen a small law firm for much of 45 years, I can appreciate this view--if the bottom line is red very many times, one quickly worries less about a sustainable society and more about a sustainable law firm. There is a dilemma here which I can define but not resolve, so I will leave it with a brief conclusion.

Help in solving environmental problems must come from the top down and the bottom up. At the top, national and religious leaders everywhere could surely do more by openly espousing effective family planning programs as well as the more gentle values already mentioned. The top can best collect and disseminate data to inform the public and show us the causative effects of our behavior. And it can lend incentives through tax policy and possibly other means. But significant improvement must move upwards from the bottom. There must first be a common perception of our predicament. We must raise social consciousness for a broad appreciation of the problems which are discussed in the final two sections of this book. The gravity of population growth and its consequences must be widely understood before a major shift in likely to occur. The media could help more in this regard. Some regions of the earth have made the demographic transition to steady or slightly declining populations, but they represent a small fraction of the world's populations and we still have a global problem. In the short run we might remind ourselves and our leaders of this problem with the chant "A million added every four days;" in the long run we must strive to make education outpace catastrophe so that we will not prove to be a species turned against itself.

References

Eisler, R. 1988. *The Chalice and the Blade.* San Francisco: Harper.

Milbrath, L. 1990. *Envisioning a Sustainable Society.* Albany, NY: State University of New York Press.

12

African Search for Solutions

Thomas R. Odhiambo

The romance of pioneering is very much the dominant spirit of Africa today. And Africans are pioneers, as they search for truly African solutions to the challenges they face. The excitement, the frustrations, the pains, and the despondency of failure all accompany the attempts of the first post-independence generation to escape the Third World and its connotation of a poverty treadmill.

In drawing inspiration from the history and philosophy of the origin of the University of Nigeria at Nsukka, I. C. Ijomah (1986) averred that "the type of education that the African needed was that education which would equip him for an independent assessment and management of his own environment," based on his own belief that "the intellect endures as potential capital in every human society." Ijomah was concerned that the African university had failed to grow as a product of its own social-cultural environment. Thus, the older universities, established during the latter part of the colonial period:

> . . . tended to isolate the undergraduate from the community in which the university was built. The mere fact that these 'fortunate few' were selected from many created the impression that they were the *creme de la creme,* and were thus entitled to be treated as the 'million dollar baby' (Ijomah 1986).

It was Ijomah's plea that the University of Nigeria at Nsukka, and by extension the new university entities established elsewhere in Africa, would search for knowledge deemed essential for national development.

Doing so, the university at Nsukka was fundamentally different from its predecessor, pre-independence universities. Indeed, Ijomah regarded the establishment at Nsukka as an essential watershed. He wrote as follows:

> For the light to shine, it must have a source. That source is the soul of Africa. To comprehend the light, we must capture the impatient, pulsating rhythm of the Africans waiting for self-realization, waiting for political independence and total emancipation from the bondage of ages. . . . The University had a mandate to prepare its products to 'dare to be free' mentally, physically, morally, spiritually, politically, economically, and socially (Ijomah 1986).

That sense of self-realization is straining to express itself in the design of a new paradigm of national development--relevant and responding directly to the needs of Africa--and not the contemporary paradigms of African development synthesized elsewhere, merely added as a footnote to a global economic order.

It is within this context that earlier chapters in this book are so illuminating, especially the three recurring themes: time as a resource; the demographic revolution; and the overarching transformational catalyst of technological innovation. Levine paints a picture of England ignoring the demographic revolution of the eighteenth century by adjusting resources to population, through "technological change [which] continuously modified the frontiers of what was humanly possible in the age-old quest to master the material world." It is his contention that the industrial revolution provided the window of opportunity through which a more democratic society could be first envisioned and then slowly realized. Such a developmental revolution has spawned new concerns, such as the continuing dilemma of juxtaposing development with environment. As Melosi summarizes so cogently "The dilemma was now how to reconcile economic benefits from industrial expansion with threats to the health and well-being of Americans and the degradation of the environment." This dilemma, so evident at the close of the nineteenth century, is still with us as we are about to close the twentieth.

Indeed, the concept of sustainable development, first made broadly known internationally by the UN's World Commission on Environment and Development, led by Gro Harlem Brundtland, in *Our Common Future* (World Commission 1986), is a dominant theme of development theory in Africa. The notion that economic development today should not diminish the prospects and options of future generations by, for example, harming the environment and reducing resource productivity, has become part of the global political agenda. Sadly, some of the key industrial factors in the green revolution of Asia and Latin America, which were to have been exported to Africa, have become doubtful technological triumphs. For instance, chemical control of insect pests, regarded as a final solution only four decades ago, has now turned out to be a "pesticide treadmill":

> On a local level chemical control could provide impressive short-term results, but over the longer term costs went up and effectiveness went down. On a larger scale, pest problems did not decrease. In fact, estimated crop losses to insect pests in the U.S. increased from about 10% in 1906 to 14% in 1974. . . . The reasons for this increase are complex, including intensification of agriculture, increased reliance of high-yielding but susceptible varieties, and expansion onto less suitable land, but a rapid increase in the use of pesticides has not been able to counteract it and has instead been a contributing factor in many cases (Kiss and Meerman 1991).

In contrast, the Wakara living on the island of Ukara in Lake Victoria, Uganda, evolved a sustainable mixed farming system for a densely populated community long before the colonial period. The system consisted of zero grazing of cattle and other domestic livestock, intercropping within a regular rotational system using both farmyard and green manure, and intensive cultivation while continuously maintaining soil productivity. The Wakara system is difficult to improve upon, even with our modern agronomic knowledge (Clayton 1964). Similarly, the Wachagga home gardens provide for both the subsistence requirements and other economic needs of large households living in a densely populated community on the slopes of Mount Kilimanjaro in Tanzania. The sustainable indigenous farming system in this case consists of mixed agroforestry, incorporating zero grazing, integrating several multipurpose trees and shrubs, and connected to a system of irrigation and drainage furrows. Fernandes et al. have no doubt about the success of the Wachagga cropping system:

> The continuous ground cover and high degree of nutrient cycling are the major factors that permit the Chagga homegardens to remain sustainable on the erosion-prone slopes of Mt. Kilimanjaro. . . . The various crop species and varieties in the homegarden represent years of natural selection for better production and quality. These species have a good resistance to prevalent pests, and compete well with weeds and have a generally high level of genetic variability (Fernandes et al. 1984)

What these two cases demonstrate is that Africa does not lack technological know-how or experience in intensive, sustainable farming under tropical conditions in order to support an increasing population. Rather, it lacks the confidence to assert so in practical terms. This lack of confidence is the greatest price that Africa has had to pay for its five centuries of diaspora and colonialism. A recent report on industrial development in Africa, from a meeting held in Nairobi, reaches a conclusion that lacks the positive message that the Wakara and Wachagga farming systems so demonstrably carry:

> The consequent retarded economic growth, inadequate infrastructure, accompanied with declines in agricultural output, and limited policy research on sustainable technologies, paint a bleak picture for sustainable industrial development in Africa. . . . Poverty and environmental degradation have been so closely interlinked that one cannot be reduced without reducing the other (African Centre for Technology Studies 1991).

Yet the poverty we are talking about here is not necessarily material poverty. It is the poverty of a mind-set determined not to learn from the experience of the very people who know their environment and their society.

Environmental debate is not a new phenomenon in Africa. Between 1786 and 1810, Mauritius provided some of the earliest experiments in systematic forest conservation, pollution control, and fisheries protection because of a fear of the possible climatic consequences of deforestation and species extinction. The pioneering environmental scientists in Mauritius, such as Pierre Poivre and Philibert Commerson, considered environmental stewardship as "a moral and aesthetic priority as well as a matter of economic necessity." The threat of induced climatic change, including a possible decline of rainfall, could lead to famine and social unrest (Grove 1990).

Two hundred years later, the African National Congress (ANC) held its first conference on science and technology in Johannesburg in November 1990. Its intent was to discuss the development of a national science and technology policy for a democratic South Africa. The ANC reached a remarkable degree of consensus. The draft position paper recognizes the need to transform the nation's technology policy from its past emphasis on defense and the oil and coal industries. Future policy should promote the kinds of applications required for a significant improvement of the living standards of its citizens and the production of goods that could compete in the international markets. Among the priority areas requiring improvement are the environmental ones: the reduction of sulfur emissions, which are serious in Eastern Transvaal; the substitution of electric power for fuel wood, in order to reduce the health hazards from wood smoke; and the development of innovative potential by creating and enabling environment for black scientists and engineers (Cherry 1990).

What Africa needs now more than anything else is the confident acquisition of a new sense of direction. We should start with the scientific rationalization of our traditional knowledge base; then ponder the range of new scientific discoveries that make new technologies appropriate to our needs possible; and then invent the social processes that make it possible for us to validate and accept these new advances.

It is evident, then, that the key to the future of the notion of sustainable development, in a world experiencing a demographic revolution in the midst of rising expectations for equity in prosperity, is *innovation* in technological terms as well as in the social sphere. This aspiration can be likened to the inspiration that a nineteenth-century Neapolitan poet evoked when he paid the following pretty compliment to the qualities of the tomato when introduced into Italian kitchens some two hundred years ago:

This dish, which is the dish
of the rich baron,
of poor folk,
of devils and saints,
of the great men of learning,
of artists and students,
this dish. . . .

(quoted in Armadei et al. 1990)

So will innovation make *Our Common Future* realizable.

References

African Centre for Technology Studies. 1991. *Sustainable Industrial Development in Africa.* Nairobi.

Armadei, G., L. Trentini, and G. P. Soress. 1990. *The Tomato.* Milan: j Enichem Agricoltura and Agrimont.

Cherry, M. 1990. ANC moves on science policy. *Nature* 348, 471.

Clayton, E. S. 1964. *Agrarian Development in Peasant Economies.* Oxford: Pergamon.

Fernandes, E. C. M., A. O. O'King'ati, and J. Maghembe. 1984. The Chagga homegardens: A multistoried agroforestry cropping system on Mt. Kilimanjaro (Northern Tanzania). *Agroforestry Systems* 2, 73-86.

Grove, R. 1990. The origins of environmentalism. *Nature* 345, 11-14.

Ijomah, I. C. 1986. The origin and philosophy of the university. In *The University of Nigeria, 1960-1985: An Experiment in Higher Education.* Nsukka, Nigeria: University of Nigeria Press.

Kiss, A., and F. Meerman. 1991. *Integrated Pest Management and African Agriculture.* Washington: The World Bank.

World Commission for Environment and Development. 1986. *Our Common Future.* New York: United Nations.

13

Transitions to a Sustainable Society

James Gustave Speth

Previous chapters in this volume have described the earliest evidence of human impact on the environment. That impact was largely local, though the case of species extinction had wider ramifications. Other chapters have traced the growing scale of impacts up to modern times. These discussions have made clear that human activities in the biosphere are now proceeding along paths that cannot be continued without dramatic reductions in the prospects for human well-being in the near future.

Acid rain, oxidants, and other consequences of fossil fuel use are affecting plant and animal life--damaging forests and fish, harming crops, changing the species composition of ecosystems--over large regions of the globe. Unsustainable agricultural practices are eroding soil, sapping water resources, and degrading rivers and estuaries.

Beyond these regional problems are planetary ones such as damage to the atmosphere and biological diversity. The community of nations has begun addressing ozone depletion, one major atmospheric threat, but not until the atmospheric concentrations of the offending chemicals had risen to levels that will produce continued damage to the ozone layer for a century to come. And, probably most serious of all, the buildup of greenhouse gases in the atmosphere continues. This buildup--largely a consequence of the use of fossil fuel and chlorofluorocarbons, deforestation, and various agricultural activities--threatens societies with far-reaching changes in climate and sea level.

Throughout the tropics, about 40 million acres of forests are being destroyed each year, roughly one acre each second. As these forests are cut down, thousands of plants and animals are lost forever. Tropical forests are the planet's richest storehouse of biological diversity, so tropical deforestation is the main force behind a species extinction rate unmatched in 65 million years, though destruction of coral reefs, wetlands, and temperate forests also play a part.

We are committing an estimated 50 species a day to extinction, and the toll could rise to one fourth of all species over the next quarter century at current rates of habitat loss. Laying waste the profusion of living things evolved over eons has profound ethical and aesthetic implications--and economic costs and as well. As plants and animal species die out, so do untold options for medical and agricultural advances that humanity will someday desperately need.

A more immediate human tragedy is unfolding in much of the developing world, where the ties between poverty and environmental degradation run in both directions. Past natural resource losses deepen today's poverty, while today's poverty makes it very hard to care for or restore the agricultural resource base and to find alternatives to deforestation. These countries are many times more dependent than industrial countries on their natural resources--soils, water, fisheries, forests, and minerals--yet this resource base is eroding rapidly, and prospects for food production and economic growth are being undermined as a result.

The U.N. Food and Agriculture Organization predicts that, without corrective action, rain-fed croplands in the Third World will become 30 percent less productive around the end of the century because the soil is depleted or eroded. Already, a billion people (mostly in South Asia and Africa) live in households too poor to obtain the food sufficient for work and normal activities. Fuelwood shortages affect an estimated 1.5 billion people in 60 countries. Ten trees are cut down for every one replanted in the Third World--30 trees for one in Africa.

The emergence of these patterns in industrial and developing countries has much to do with the successes and failures of economic activity. The twentieth century has witnessed explosive growth: since 1950, world population has doubled to over five billion, and the world economy has quadrupled. One result is that pollution and waste generation are occurring on a vast and unprecedented scale. Global fossil fuel use has increased ten-fold in this century, and the resulting emissions have grown dramatically. Human demands on biological systems have increased to the point that we consume an estimated 40 percent of the world's total terrestrial photosynthetic productivity, and much of this is occurring in a way that is not biologically sustainable. For the first time, human impacts have grown to approximate those of the natural processes that control the global life-support system.

These challenges are closely interlinked, planetary in scale, and deadly serious. They cut across sectors and regions. They cannot be addressed issue-by-issue or by one nation or even by a small group of nations acting alone. They will not yield to modest efforts in the face of a possible further doubling of world population and quintupling of world economic activity in the lifetimes of today's children.

To address these challenges, a series of large scale social and economic transitions is needed. Everywhere, environmental deterioration is integrally related to economic production, technology, the size of human and animal populations, social equity, and a host of other factors. This deterioration is unlikely to be reversed except through broad macro-transitions with multiple social benefits. These transitions are needed in both industrial and developing countries, and cannot be achieved without a genuine partnership between North and South.

These transition are also far different from the environmental protection approaches of the early 1970s. Environmentalism began on the periphery of the economy, bottling up pollution here, saving a piece of landscape there. Now, it must spread as creed and code to permeate to the core of economic activity. It must deal with the root causes of environmental problems, recognizing that the solutions to underlying causes lie mostly outside the established "environmental sector." The changes that are essential for treating the root causes are encapsulated in the following six transitions.

A Demographic Transition

Transition One is a demographic transition toward stable populations, both in nations where growth is explosive and on a global basis, before the world's population doubles again. Population size and its growth are contributing factors to virtually every environmental challenge societies face today. It is hard enough to imagine a workable world of ten billion people, twice today's level, but consider that the most recent U.N. projection puts global population at 13 billion in the next century, if fertility does not fall rapidly.

Since World War II, the industrial world's birth rates have fallen toward--in places dipped below--replacement level, as social and economic prospects improved. Now it is critical that developing countries make a similar demographic transition, and raising living standards in a necessary condition. Also essential are improving the status of women, expanding their access to education and income, improved sanitation and health care to reduce infant mortality rates, making family planning services universally available, and providing some kind of support for the elderly. Developing countries that set out to reduce population growth rates must also be able to count on help from the wealthy nations.

A Technological Transition

It is also necessary for the world to move away from today's resource-intensive, pollution-prone technologies to a new generation of environmentally benign ones. We need a worldwide environmental revolution in technology--a rapid, ecologically sound modernization of industry and agriculture. The only way to reduce pollution and waste while achieving expected economic growth is to transform the dominant technologies of manufacturing, energy, transportation, building design, and agriculture. The nineteenth-and twentieth-century technologies that have contributed so abundantly to today's problems must be rapidly replaced with twenty-first-century technologies that dramatically reduce environmental impact per unit of output.

Nowhere is this transition more urgently required than in the energy sector, given the side effects of today's extravagant fossil fuel use. We must begin with efficiency gains, promoting the widespread use of cost-effective technologies that markedly reduce the energy needed in industry, homes, and transportation. But such gains will eventually be canceled out by growth unless alternatives stand ready. Solar thermal and wind systems already produce electricity at prices competitive with nuclear power plants, and photovoltaics may do so by the decade's end. In the real world of transportation, cars and trucks are a prime target for change. Besides stepping up gas mileage immediately, we will eventually need mass-marketed "green cars" powered by hydrogen or electricity made from renewable, non-polluting energy sources.

Markets and regulations must be made to work for this transformation in technology, not against it, as commonly happens today. The private sector, "green" consumers, and countless local, national, and international organizations must be as fully engaged in this transition as governments. Wealthy nations must support the transfer of solution-oriented technologies on affordable terms to the former communist states of Eastern Europe and to developing countries.

An Economic Transition

The world economy must be based on reliance on nature's "income" and not depletion of its capital. In countries around the world, inappropriate subsidies and other policies encourage deforestation, excess pesticide use, energy inefficiency, and water wastage. National income accounts that treated natural resource deterioration correctly--as another form of capital depreciation--would show that today's societies are living off nature's capital to an alarming degree.

In market economies, this economic transition depends on environmentally honest prices--prices that include the full environmental costs of production, use, and disposal. It has been said that the planned economies failed in part because prices did not reflect *economic* realities. It might also be said that the market economies will fail unless prices reflect *ecological* realities.

Accurate pricing will require measures such as pollution taxes, user fees on virgin materials, and eliminating subsidies. Because such measures can also enhance government revenues, the needed revolution in pricing can go hand in hand with shifting of a share of the tax burden away from activities that should be encouraged, such as earning income and investing, and onto activities that should be discouraged, such as generating pollution and waste.

A Social Transition

We also need a social transition to more equitable sharing of environmental and economic benefits, among and within nations. Today, the wealthy consume inordinate quantities of the world's natural resources, while the poor have little choice but to overtax the resource base that must sustain them tomorrow. Over much of the developing world, poverty is both a cause and an effect of environmental degradation. Developing countries desperately need major new financial resources dedicated to sustainable development. Sharply increased North-to-South flows of official development assistance and private investment must be accompanied by improvements in the ways that both existing and added funds are spent. Channeling these funds to generate environmentally sustainable employment opportunities for the poor is crucial to this transition. So are raising the status of women, reforming land tenure, strengthening nongovernmental organizations, nurturing the growth of democracy, promoting demilitarization, and stemming corruption.

Eradicating poverty--essential both to meet human needs and to take pressure off a deteriorating resource base--is unlikely unless the industrial countries increase development assistance, relieve the burden of international debt, and reduce their trade barriers against labor-intensive imports. The wealthy North should also be setting an example by amending its own unsustainable development practices (for instance, by conserving old-growth forests and curbing fossil fuel use) and lifting its own poor people out of poverty.

A Transition in Consciousness

A more profound and widespread understanding of global sustainability is needed. Our understanding of natural systems and human impacts on them is deepening, but it is still no match for nature's complexities or the often subtle connections between one environmental problem and another. More attention must be paid to monitoring environmental conditions, assessing trends, and devising indicators that can help both policy makers and the public grasp important trends and gauge progress. We need more scientific research to reduce uncertainties on global warming and other planetary changes. Since setting the right fiscal, economic, and resource policies is key to improving environmental conditions anywhere, we need more policy research to identify the best means of relieving pressures on the earth's resources. Funding for both lines of inquiry must be viewed as an investment in creating knowledge that will more than pay for itself. Publicizing research findings as widely as possible and improving environmental education are critical since even the most startling discoveries cannot be translated into change without widespread public understanding and support. Equally important is the need for professional and other training in environmental management--building capacity that is in desperately short supply today.

An Institutional Transition

Also needed is a shift to new arrangements among governments and peoples that can achieve environmental security. Because environmental policy will increasingly be set in concert with other nations, the United Nations and other institutions need the capability to reach a broad array of international agreements more swiftly and surely and with less *ad hoc* activity. Because environmental objectives cannot be met through environmental action alone, new arrangements are needed to integrate these objectives into such other fields as trade and debt policy, agriculture, energy, transportation, foreign policy, and development policy and assistance. Because none of today's major environmental challenges can be met without a new era of heightened cooperation between industrial and developing countries, institutional innovations are needed to facilitate over time a complex set of understandings and agreements--a compact or bargain--between North and South.

This institutional transition is needed at all levels, from the top down and from the bottom up. For world political leaders, the challenge is to devise a new system of shared international responsibility. The principal goal of diplomacy must shift from conflict management to common endeavor. Equally significant for this next round, though, will be the activity of private business, citizen groups, nongovernmental organizations, and communities. In both industrial and developing countries, countless local and private sector initiatives are a potent and growing force for change. They need encouragement, support, and in many cases funding from development agencies.

Priorities for the 1990s--and Beyond

With the fate of humanity and nature written in policies that nations everywhere adopt over the next decade, vision and political will must become every nation's most valuable assets. World leaders should agree that a priority mission of international cooperation and diplomacy during this decade will be revamping the patterns and practices that have brought us to this pass. And they should adopt specific initiatives that will promote these essential transitions to a sustainable human society.

The Four Horsemen of the modern age have been the Cold War and the arsenals it has spawned; widespread suppression of human rights; global poverty and hunger; and unrelenting assault on the environment. Today, the first two Horsemen seem to be in retreat, but the last two loom larger than ever. As East-West tension winds down and democracy spreads, we must both encourage these recent geopolitical trends and launch in earnest a global campaign against poverty and environmental deterioration.

Notes

This paper is based on the Annual Hagood Lecture given by the author at the University of South Carolina School of Law on February 27, 1992.

Data reflecting all aspects of environmental change can be found in:

World Resources 1992-93. 1992. A report by the World Resources Institute in collaboration with the United Nations Environment Programme and the United Nations Development Programme. New York and Oxford: Oxford University Press.

About the Contributors

Mark N. Cohen is professor of anthropology at the State University of New York at Plattsburg. His latest book is *Health and the Rise of Civilization* (1989).

John Firor is director of the Advanced Study Program at the National Center for Atmospheric Research and author of *The Changing Atmosphere: A Global Challenge* (1990).

Peter H. Gleick is director of the Global Environment Program of the Pacific Institute for Studies in Development, Environment, and Security. He is the editor of the forthcoming book *Water in Crisis: A Guide to the World's Fresh Water Resources* (1992).

C. M. Hudspeth is a lawyer in Houston, Texas, and serves as counsel to the firm of De Lange, Hudspeth and Pitman.

Judith E. Jacobsen is assistant professor in the Department of Geography and Recreation of the University of Wyoming and is a former senior researcher at the Worldwatch Institute, Washington, D.C.

Richard G. Klein is professor of anthropology at the University of Chicago. He has studied the ecology of paleolithic people in Africa and Europe and is author of *The Human Career: Human Biological and Cultural Origins* (1989).

David Levine is professor of history and philosophy at the Ontario Institute for Studies in Education. He is the author of *Reproducing Families: The Political Economy of English Population History* (1987) and other books on the demographic implications of industrialization.

Martin V. Melosi is professor of history at the University of Houston and author of several studies on the urban environment and energy history including *Coping with Abundance: Energy and Environment in Industrial America* (1985).

James D. Nations is vice president for Latin American programs of Conservation International.

William A. Nitze is president of the Alliance to Save Energy, Washington, D.C., and a former deputy assistant secretary of state for Environment, Health and Natural Resources. He is the author of *The*

Greenhouse Effect: Formulating a Convention (1990) and papers and articles on energy and environmental issues.

Thomas R. Odhiambo is director of the International Centre of Insect Physiology and Ecology at the University of Nairobi, Kenya, and president of the African Academy of Science.

Charles L. Redman is professor in the Department of Anthropology of Arizona State University. Professor Redman has excavated in the Near East, North Africa, and the American Southwest and is author of *The Rise of Civilization*.

James Gustave Speth is president of the World Resources Institute in Washington, D.C., and former chairman of the U.S. Council on Environmental Quality.

G. N. von Tunzelmann is reader in the economics of science and technology at the Science Policy Research Unit (SPRU) at the University of Sussex. He is the author of *Steam Power and British Industrialization*.

Index

Acid rain, 207
Afforestation, 184
African National Congress, 204
Agricultural economy
  and animal populations, 39
  definition, 36
  diversity vs. specialization, 40
  and human organization, 40
  and natural landscape, 39
  and political organization, 40
  and population aggregation, 39
  and population growth, 38
  and soil erosion, 41
  See also Sustainable agriculture
Agroforestry, 203
American Southwest, 42-45, 56
Anasazi, 44-45
Aquifers, 197

Biological diversity, 172, 207
  See also Tropical forests
Biotechnology, 189
Bottom up, 184, 199, 213

Chipko movement, 194
Chlorofluorocarbons, 207
Climate change, global, 15, 42-44, 143, 153, 183
  societal impact of, 153
  and water resources, 165
Climate model, 148
  verification techniques, 151

Coal
  and British industry, 135-136
  cost and output, historical, 120
  and cost reductions, 127
  costs of production, historical, 119-126
  demand price, historical, 125
  diminishing returns, 118, 122-126, 129, 132, 136
  exhaustibility, 115, 117, 126
  and industry organization, 131-133
  and new discoveries, 129, 136
  patents, 129-130, 132
  physical depletion vs. cost 116
  price, historical, 123
  production volume, historical, 132
  real costs in terms of labor, historical, 124
  and Reid Committee of 1945, 132
  and safety standards, 128
  and Samuel Commission of 1925, 131
  and strikes, 127
  and substitutes, 134-136
  and technological innovation, 129-132, 136
  technology, 125
  use efficiency, 132-134
  and wages, 127-128, 132, 136-137
  working costs, 123
Codex, Mayan, 171

Deforestation, 41, 204, 207-208
deLanda, Bishop Diego, 171

Deming, W. Edwards, 188
Drift nets, 198

Ecuador, 173, 178
Ehrlich, Paul, 119
Energy efficiency, 40, 62, 96,
    101-102, 110, 184
Environmental externalities, 191
Environmental reform
  by citizen organizations, 101
  the courts, 103
  education, 102
  government regulation, 104-109
  legislation, 107-109
  by professionals, 101
Extinction, large mammal, 18-21
  and the fossil record, 18-19
  and climatic and environmental
    change, 22, 26-27
  history, 19-21
  and modern humans, 23-27
  and premodern humans, 21-22

Family, 53-54, 81-85
  and sexuality, 81
  Malthusian couple, 83
  policing of, 83
  and private rights, 81
  sentimentalization of, 85
Fertility decline, 73-75, 77, 116-117
  and national differences, 77
  and spacing pattern, 77
  and starting pattern, 77
  and stopping pattern, 77
Fossil fuels, 91-93, 104, 109-110,
    207
Fossil fuel burning, 147
Four Horsemen, 213

Greece, Ancient, 41-42
Gross National Product, 190

Guatemala, 171, 176-178

Heat trapping gases
  carbon dioxide, 144
  change of, 156
  concentration of, 156
  halocarbons, 144-145
  methane, 144-145, 148
  nitrous oxide, 144-145
  stabilization of, 156
  water vapor, 144, 146
Hohokam, 43-44
Hotelling, H., 117-118, 126, 136
Human civilization, 18, 35-40, 45,
    48, 171
  definition, 53-54
Human evolution, 14-18

Infectious disease
  AIDS, 64
  and American Indians, 66
  and animal domestication, 60-61
  bubonic plague, 56, 60, 64
  cholera, 64
  chronic disease, 55-56
  common cold, 61
  dengue fever, 57, 64
  and diet, 62
  and early human populations, 55
  epidemic disease, 56-57, 65
  Epstein-Barr, 55, 63
  fecal-oral diseases, 58-59
  herpes, 55
  hookworm, 59
  and human behavior, 52-53, 175
  and human impact on vector and
    reservoir species, 57-58
  and hunter-foragers, 59, 62
  and hygiene, 62-63
  influenza, 60-61, 64
  Lyme disease, 57
  malaria, 57, 59, 63-64
  measles, 61, 63, 66

mumps, 63
nature of, 51-52
osteomyelitis, 55
plague, 63
polio, 63
and political history, 65-66
and population density, 54-55
rabies, 56, 60
Rocky Mountain spotted fever, 57
schistosomiasis, 58, 64
and sedentism, 58-60
small group epidemiology, 55-56, 65
smallpox, 61, 63
syphilis, 63-64
tapeworm, 61
tetanus, 61
tick-borne diseases, 57, 60
toxoplasmosis, 60
and trade, 63
tuberculosis, 55, 61, 63
typhus, 60, 63
as a weapon, 66
and the Yanomamo, 66
yaws, 55, 63
yellow fever, 57, 63-64
zoonoses, 56

Jevons, William Stanley,
The Coal Question (1865), 115

Malthus, Thomas, 116
Malthusian population theory, 115, 117
Malthus-Ricardo theories of rent, 115
Mauritius, 204
Maya, 171
Mesopotamia, 45-48
Migration, 75
as cause of disease, 64
rural-to-urban, 98
Montreal Protocol, 185
Mortality decline, 74-75
and migration, 75
and age pyramid, 75

Nagymoros Dam, 194
Natural resources
value of, 192
Neoclassical economies, 118

Oil pollution, 94-96
in air, 96
in Pennsylvania, 95
in Texas, 96
in water, 96
OPEC, 118
Oxidants, 207
Ozone depletion, 207

Pesticide treadmill, 203
Pollution
as barbarity, 101
as health hazard, 100
as nuisance, 100
as waste, 100
See also Oil pollution, Water pollution
Pollution remedies
court action, 99, 102
education, 99, 102
legislation, 99
protests, 99-100
regulation, 99, 104
Proletarian proliferation, 78
and individualism, 79
and Malthus, 79

Renewable and non-renewable resources, 92, 117, 163

Salinization, 44, 46-48, 164
Simon, Julian, 118-119, 124
Species extinction, 13-14, 204, 208
State formation, 54, 74, 78
  and the classroom, 78
  and nationalism, 78
  and sex, 78
Stratospheric ozone destruction, 143
Surface temperature of the earth, 147
  heating of, 150
Sustainable agriculture, 194
Sustainable development, 202

Technological change and energy consumption, 76, 99
Thiobacillus ferrodoxins, 189
Top down, 184, 199, 213
Top soils, 35, 37, 41-42, 56, 197
Tropical forests
  and anti-cancer drugs, 177
  and beef cattle production, 174
  biological diversity, 172
  and cash crops, 174
  causes of migration to, 173-174
  Chamadoria *(xate)* palm, 177-178
  and chicle, 177
  colonization, 173
  and cortisone, 177
  and curare, 177
  destruction, 173
  economic arguments, 178
  and global climate change, 179
  and indigenous peoples, 177-178
  Latin America, 173
  and non-food materials, 1177
  rain forests, 173
  roads, 173
  as source of common foods, 175
  as source of crop germplasm, 175-176
  as source of new foods, 175-176
  spiritual values, 178
  sustainable use, 172, 177-179
  tagua, 178
  and tourism, 178
  value, 172, 174-177
  Xateros, 178

United Nations Environment Program, 186
Urbanization, 39-40, 53-55, 58-59, 64, 73-76, 98
  and proletarianization, 74

Wachagga, 203
Wakara, 203
Water pollution
  arsenic, 97
  benzene, 97
  and commercial fishing, 97
  sewage, 97
Water resources, global, 161-162
  California drought, 166-167
  competing demands for, 164
  and energy, 165
  Euphrates River, 168-169
  and food production, 37-38, 43, 47, 164
  and global climate change, 165-167
  and international conflict, 167-169
  irrigation, 43-45, 46-48, 164, 167
  maldistribution, 162
  Nile River, 167-168
  and population growth, 164, 167
  safe drinking water, 164
  and sustainable development, 45, 163-165
Water use, 37-38, 43, 46-47, 58, 163
World population, 38, 41-42, 74, 97-98, 148, 187

Yanomamo, 66
Yucatan, 172